D0215128

The Basis and Applications of Heterogeneous Catalysis

Michael Bowker
Department of Chemistry, University of Reading

Series sponsor: **ZENECA**

ZENECA is a major international company active in four main areas of business: Pharmaceuticals, Agrochemicals and Seeds, Specialty Chemicals, and Biological Products.

ZENECA's skill and innovative ideas in organic chemistry and bioscience creates products and services which improve the world's health, nutrition, environment and quality of life.

ZENECA is committed to the support of education in chemistry and chemical engineering.

OXFORD NEW YORK TOKYO
OXFORD UNIVERSITY PRESS
1998

Oxford University Press, Great Clarendon Street, Oxford OX2 6DP

Oxford New York
Athens Auckland Bangkok Bogota Bombay
Buenos Aires Calcutta Cape Town Dar es Salaam
Delhi Florence Hong Kong Istanbul Karachi
Kuala Lumpur Madras Madrid Melbourne
Mexico City Nairobi Paris Singapore
Taipei Tokyo Toronto Warsaw
and associated companies in
Berlin Ibadan

Oxford is a registered trade mark of Oxford University Press

Published in the United States by
Oxford University Press Inc., New York

First published 1998

A catalogue record for this book is available from the British Library

Library of Congress Cataloging in Publication Data
Bowker, M. (Michael)
The basis and applications of heterogeneous catalysis / M. Bowker.
(Oxford chemistry primers; 53)
Includes bibliographical references and index.
1. Heterogeneous catalysis. I. Title. II. Series.
QD505.B694 1998 541.3'95—dc21 97–42914
ISBN 0 19 855958 5

Typeset by Author
Printed in Great Britain by
The Bath Press, Bath

Series Editor's Foreword

Oxford Chemistry Primers are designed to provide clear and concise introductions to a wide range of topics that may be encountered by chemistry students as they progress from the freshman stage through to graduation. The Physical Chemistry series aims to contain books easily recognized as relating to established fundamental core material that all chemists need to know, as well as books reflecting new directions and research trends in the subject, thereby anticipating (and perhaps encouraging) the evolution of modern undergraduate courses.

In this Physical Chemistry Primer Professor Michael Bowker presents an authoritative, clearly written, and elegant account of the fundamentals, practicalities, and applications of *Heterogeneous Catalysis*. This is a topic of immense industrial and environmental significance so that this Primer will be of interest to all students of chemistry and their mentors.

Richard G. Compton
Physical and Theoretical Chemistry Laboratory
University of Oxford

Preface

Heterogeneous catalysis is one of the most important technologies for our modern society. It underpins the chemical and materials industries, being fundamental to the production of fuels and plastics, for instance. This strategic importance has recently been stressed in reports to both American and British governments, resulting in specific support for this area of science. Thus, it is essential that a Primer covers this extensive subject. The book is divided into three sections dealing with fundamental aspects of the phenomenon, the nature of catalytic materials and the applications of catalytic technology in everyday modern life. A very important and expanding area of application is in helping to solve pollution problems and this is addressed in Chapter 8. The book is aimed at undergraduate students and their teachers who may be involved with individual courses in this field. It will also be useful, however, to Master's students and those beginning a research career in catalysis, surface science, and related fields. It covers most of the main themes important in catalysis, namely the nature and kinetics of adsorption and surface reactions, catalyst preparation and characterization, and the application of these ideas to the needs of society for particular products and processes.

Finally, I would like to express thanks to several people for their involvement with this book. First and foremost, my student Richard Holroyd who organized the text and figures through several drafts; his uncomplaining hard work has helped see the book through to completion. I am also grateful to a number of people who read the manuscript; thanks are especially due to Dr Mark Howard of BP Chemicals Ltd, Hull and to Prof. Jack Frost of Johnson Matthey Technology Centre, Sonning Common for their helpful comments.

Reading M.B.
November 1997

Contents

I Fundamentals: what is catalysis?

Catalysis is one of the most important technologies in our modern world. We depend on it to produce materials, such as plastics, from oil; we depend on it to produce fuel to power our cars; we depend on it to remove the pollutants emitted from the engines of those cars; we even depend on it for the functioning and growth of our own bodies. It is therefore very important that we ask ourselves the question, 'what is catalysis?' In this section the fundamental basis of catalysis will be described, focusing largely on the crucial part of a catalyst, namely the surface layer, where all the chemical action takes place.

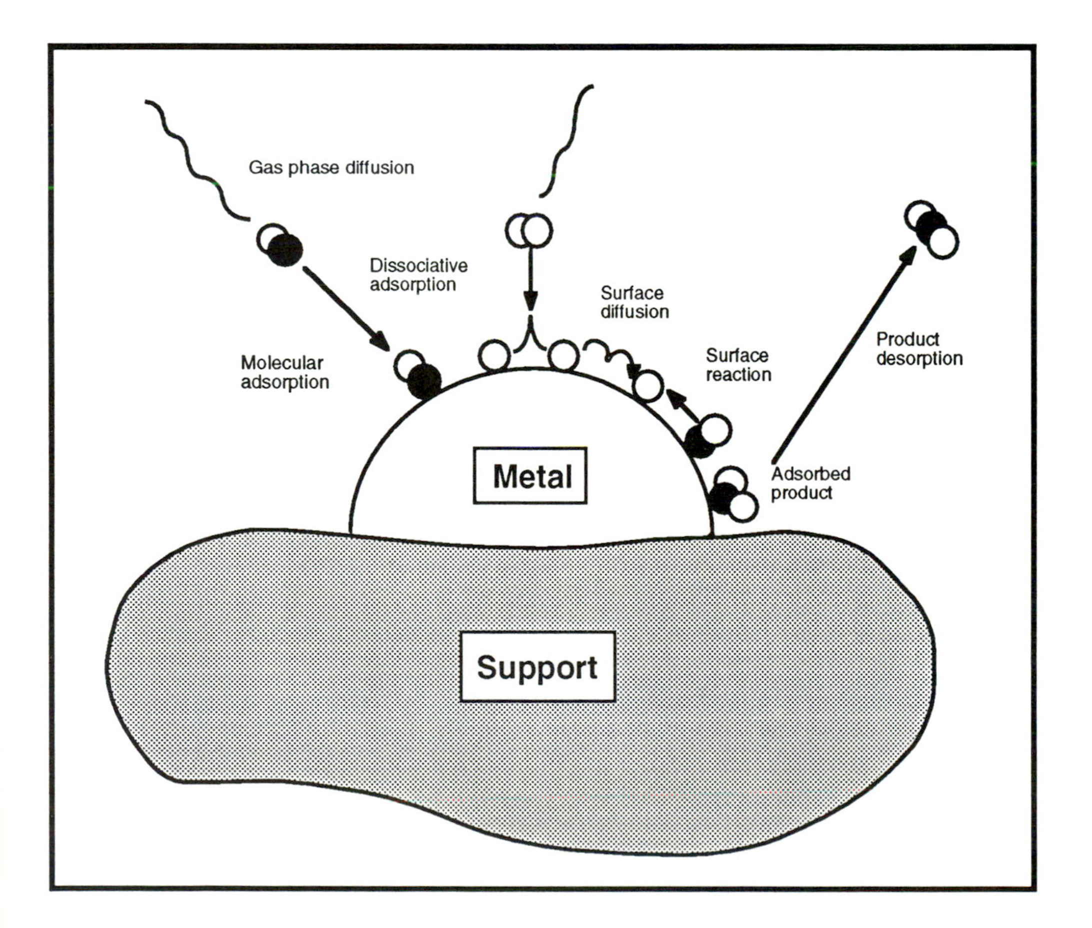

1 The reactive interface

1.1 Background

Catalysis is an extremely important phenomenon for our modern industrial economy, without it life today would be quite different from the reality we see around us. As one example of this, and perhaps the most widespread application of catalysis today, cars are fitted with catalytic converters to reduce the pollutants (such as NO) leaving the vehicle. There is no doubt that without this technological advance, our cities would be much more severely polluted places than they currently are. Catalysis is a vital technology in today's world. Approximately 90% of all our chemicals and materials are produced using catalysis at one stage or another. In turn, many developed countries are net exporters of such chemicals and thus rely on catalysis for the health of their economies.

Catalysis is not a new phenomenon, although its wide-scale application by mankind has only really begun in this century. Enzymatic catalysis is essential to all living matter, without it life would not exist. Most fundamental of all of these processes, perhaps, is photosynthesis, a process for energy storage which is used by most of the simplest and earliest evolved lifeforms. After photon capture in chlorophyll and storage of the energy as separated charge, several enzymes operate in the production of sugars, which is the chemical form in which energy is stored, and which is ultimately also used by humans.

Humans learned to utilize natural catalytic processes very early on, exactly how early is difficult to say, but natural yeasts were used to ferment fruits to yield alcoholic drinks many millennia ago. Indeed, ancient Sumerian physicians prescribed a kind of beer as a 'pick-me-up' for their patients; the prescriptions were written on clay tablets. The natural catalysts in yeast which convert sugar into alcohol are enzymes, and a number are involved in this complex, step by step conversion.

'Many bodies have the property of exerting on other bodies an action which is very different from chemical affinity. By means of this action they produce decomposition in bodies, and form new compounds into the composition of which they do not enter. This new power, hitherto unknown, I shall call it catalytic power. I shall also call catalysis the decomposition of bodies by this force.' J. J. Berzelius, Edinburgh New Philosophical Journal, 21 (1836), 223.

A more systematic study of the phenomenon began early in the 19th century. Studies of this process began to accelerate, with a number of notable scientists being involved. Davy began experiments in 1815 on catalytic combustion with Pt gauzes (Pt having then only recently been discovered and extracted). Indeed, the catalytic oxidation of 'fire-damp' on the gauze in the Davy safety lamp (which was designed for use in mines), warned miners that they were in a dangerous area since the gauze would glow in such a gas.

A little later in the century, Berzelius coined the term 'catalysis' after working in this field of chemistry for a number of years. The quotation shown in the margin gives us a good idea of what catalysis is about, but over the years there have been some misunderstandings regarding the nature of catalysis. A catalyst does not appear in the stoichiometric equation for an overall reaction, but it is nevertheless directly involved in the conversion and appears both in individual mechanistic steps, and in the kinetic rate law. A simple example of homogeneous catalysis is ozone destruction, a reaction of importance to the ozone hole phenomenon, as described in the margin.

A slightly simplified and updated definition of catalysis might be as follows – 'Catalysis is a process whereby a reaction occurs faster than the uncatalysed reaction, the reaction being accelerated by the presence of a catalyst'. Note that acceleration of an elementary reaction step is produced by the catalyst and that at equilibrium (equal rate of forward and reverse reactions) no gain is achieved by the addition of a catalyst – under such circumstances the reaction is said to be under 'thermodynamic control' and catalysts are only effective when a reaction is under 'kinetic control'. This will be discussed in more detail later (Section 6.1). The definition of a catalyst, then, is linked to the definition above, and is given in the margin.

An example of simple catalysis

$O_3 + Cl \rightarrow ClO + O_2$
$ClO + O \rightarrow Cl + O_2$
Overall reaction
$O + O_3 \rightarrow 2O_2$
Here, ozone destruction is catalysed by Cl radical atoms, the Cl is directly involved in the reaction, and an intermediate with Cl in it is formed. Note, however, that Cl is regenerated at the end of the catalytic cycle and that it does not appear in the overall stoichiometric equation.

In the late nineteenth century the application of catalysis accelerated and Table 1.1 shows some notable dates for the development of a few large-scale industrial processes. Perhaps the most well known of these is the Haber synthesis for ammonia production which was developed in Germany just prior to World War I, thus ensuring supplies of fertilizer which had previously been imported from natural sources in South America. This technology has now been applied world-wide with little modification to the mainly Fe catalyst which was developed then.

A catalyst is a body or a material which can induce the phenomenon of catalysis. It enhances the rate of the catalysed reaction, and while being intimately involved in the reaction sequence, it is regenerated at the end of it.

In more recent times catalysis has been applied to national defence, for example, to the security of the UK during World War II. The Spitfire fighter aircraft has an heroic place of honour, together with the men that flew them, in the annals of World War II history, particularly in the 'Battle of Britain'. What is less well known is that the application of catalytic innovation was also behind their success. The Spitfires were able to use a new, higher octane fuel derived from catalytic oil cracking technology (see Chapter 7), which helped give them better acceleration than their opponent aircraft.

Table 1.1 Early large-scale catalytic processes

Reaction (Discoverer)	Catalyst (Date)
$2HCl+\frac{1}{2}O_2 \rightarrow H_2O+Cl_2$ (Deacon)	Copper (~1860)
$SO_2+\frac{1}{2}O_2 \rightarrow SO_3$ (Phillips)	Platinum (1875)
$CH_4+H_2O \rightarrow CO+3H_2$ (Mond)	Nickel (1888)
$2NH_3+\frac{5}{2}O_2 \rightarrow 2NO+3H_2O$ (Ostwald)	Pt foil (1901)
$C_2H_4+H_2 \rightarrow C_2H_6$ (Sabatier)	Pt (1902)
$N_2+3H_2 \rightarrow 2NH_3$ (Haber)	Promoted Fe (~1914)

Our present economy is highly geared to the use of catalysis. We produce fuel for our cars by cracking and reforming oil into petrol and diesel using a variety of catalysts. The ethylene and propylene from the cracking of naphtha is used in large-scale polymer production, and we even modify our food using catalysis (spreadable margarine is produced by saturation of the double bonds of unsaturated, free-flowing oils using nickel catalysts). Practically every aspect of modern living is dependent on catalysis and so it is a subject well worth study and understanding.

This small book is an attempt to spread the understanding of catalysis, and in the next section we begin this process by considering the nature of the surfaces of materials, since this is where heterogeneous catalysis takes place. The nature of the topmost atomic layer which is exposed to the reactant is crucial to the efficiency of all such processes.

1.2 Surfaces are crucial for catalysis

Heterogeneous catalysis – the catalyst and reactant are in different phases, e.g. ammonia synthesis from N_2 and H_2 over a solid Fe catalyst.

Homogeneous catalysis – the catalyst and reactants are in the same phase, e.g. O_3 and Cl in the atmosphere as described previously.

Industrial catalysis can be divided into two broad types – heterogeneous and homogeneous. Most large-scale, industrially catalysed processes are of the former type, and the widespread recent application of catalysis to car emission control uses such solid catalysts in contact with the gas phase exhaust stream. For these kinds of reaction the nature of the interface is crucial for the efficiency of the process. The nature of the top layer of atoms determines how fast a catalytic reaction takes place and small amounts of additives can reduce (poison) or enhance (promote) the reaction.

Poison – usually blocks active sites on the surface of the catalyst, thus reducing reaction rate. Usually electron acceptor elements such as Cl, S, C.

Promoter – enhances the activity of the catalyst by improving reactions at sites adjacent to the promoter. Usually an electron donor element, e.g. K, Cs, La.

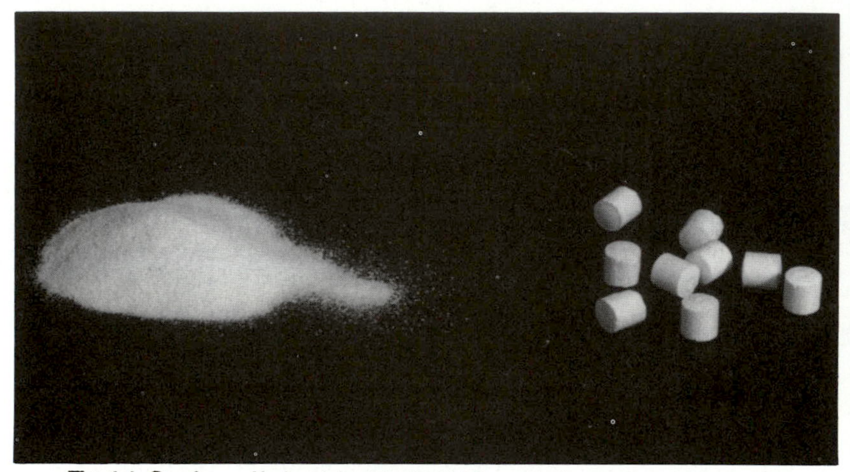

Fig. 1.1 One form of industrial catalyst which is made from compacted powder.

Figure 1.1 shows the typical form in which an industrial catalyst is used-a pellet. The geometrical area that is apparent is only approx. 1 cm^2, but the material is highly porous having a network of small pores within its structure. The pellet is manufactured from a fine powder which is compressed into the final form. Figure 1.2 shows an example of what this powder looks like at the microscopic level, imaged using scanning electron microscopy (SEM). This is a $Ag/\alpha\text{-}Al_2O_3$ catalyst which is used for ethylene epoxidation catalysis (Eqn 1.1), this being the largest industrial selective oxidation process in operation, the product being the precursor chemical used to make antifreeze, washing powder whiteners and a wide range of other products.

$$C_2H_4 + \frac{1}{2}O_2 \rightarrow C_2H_4O \qquad (1.1)$$

The SEM picture in Fig. 1.2 shows small particles of Ag (approximately 1 μm in diameter) bonded to the support material, $\alpha\text{-}Al_2O_3$. The support material is there in order to maintain the integrity of the metal phase, where the selective reaction takes place. The area of the Ag is crucial for efficient catalysis. Without the support, the metal would undergo sintering (fusion of the particles) very rapidly and would therefore lose activity.

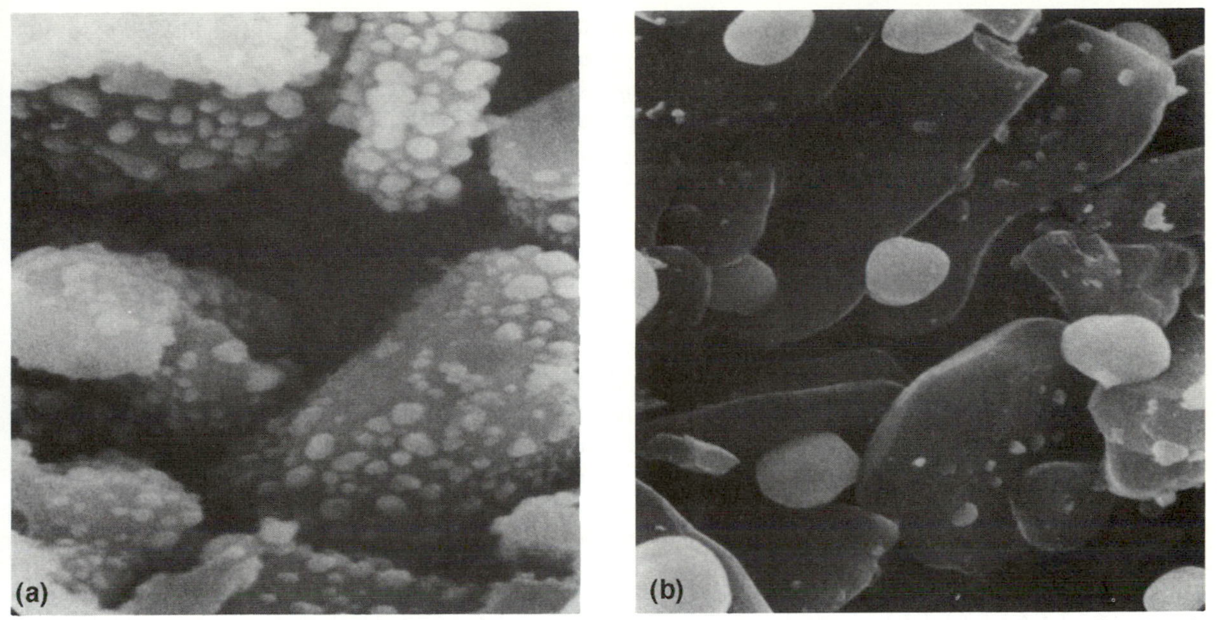

Fig. 1.2 Scanning electron micrographs of a $Ag/\alpha\text{-}Al_2O_3$ catalyst scanning (a) before and (b) after use in a microreactor. Sintering of the metal particles has occurred (sintering is discussed further in Section 2.2). Scale of figure is 10 μm.

It is not only the amount of this surface area, but also its detailed composition and structure, that is important for the efficiency of the reaction. An example of what the surface of a catalyst particle shown in Fig. 1.2 might be like at the atomic scale is shown in Fig. 1.3. This is an image of a very small diameter (approx. 100 nm) Rh hemisphere taken with a special technique called field ion microscopy (FIM) and each bright spot is an atom on the surface of the particle. It is evident that different morphologies are exposed: some flat planes of atoms, some steps, even missing atoms. These atoms all have different coordination at the surface and this affects their ability to bind and react with gas phase molecules.

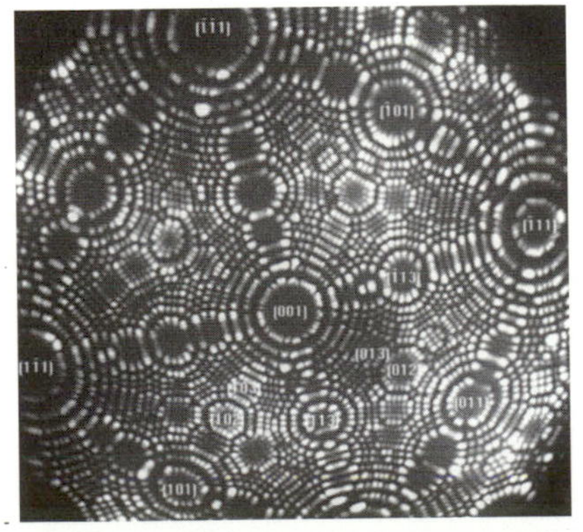

Fig. 1.3 Field ion micrograph of a (001)-oriented rhodium specimen, showing individual atoms at the surface. (Courtesy of Prof. N. Kruse, Universite Libre de Bruxelles.)

This low coordination is the essence of catalysis, and is the driving force for it. The surface is an abrupt termination of the bulk structure and so exposes atoms in an asymmetric environment, with neighbouring atoms only in the surface plane and in the bulk direction. This leaves free bonds at the surface which are available for interaction with incoming molecules. This and the subsequent steps of catalysis are discussed in more detail in Section 1.3 below.

The driving force for catalysis then, is these free bonding states at the surface, and, in thermodynamic terms, this 'catalytic power' is represented by the surface free energy, G_S.

Because of the termination and asymmetry the surface region is one of high energy, E_S. This energy can be understood in terms of the work it requires to make a surface (Fig. 1.4). In real terms this can be felt as the muscle power required to cut through an aluminium bar, or the effort needed to tear a metal foil into two halves, both of these break metal bonds and create new surfaces. The former requires much more effort than the latter because the area of surface created is much higher and more bonds are broken in the process.

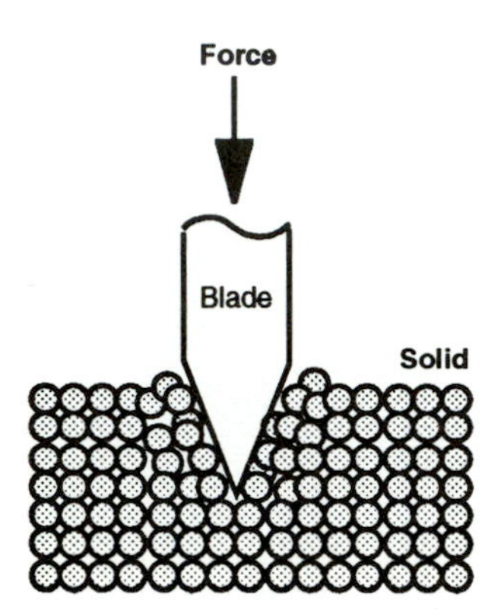

Fig. 1.4 Work is needed to break bonds in a solid, in this case cutting a solid with a sharp blade.

1.3 The catalytic cycle

The molecular level processes involved in product formation are illustrated in Fig. 1.5 for a very simple example, CO oxidation on a metal surface, which is nevertheless practically important because this reaction takes place in automobile exhaust catalysts. The first step is diffusion of the molecules through the gas phase to the metal surface where the molecules may bond (adsorption) in a molecular form.

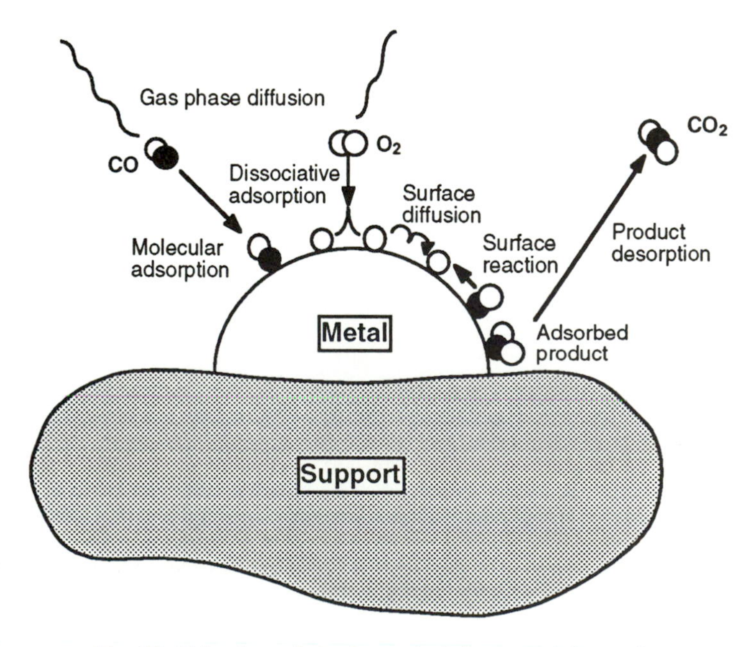

Fig. 1.5 Molecular and atomic events during a catalytic reaction.

Surface diffusion may then occur, and the molecule may dissociate into atoms. For the particular example shown, O_2 dissociates, but CO does not, and this is because of the much higher internal bond strength in CO (1076 kJ mol^{-1} vs. 500 kJ mol^{-1} for O_2). The next step is a surface reaction whereby oxygen atoms react with the CO to form the adsorbed product, CO_2. The surface reaction step is often the rate-determining step in a catalytic reaction. Finally, product desorption occurs, where the product to surface bond is broken and the CO_2 enters the gas phase, diffuses down the catalyst pores, to emerge finally at the end of the reactor. A great deal of engineering expertise is required in catalytic processes in order to manipulate gas phase diffusion. This is because very often the product carries excess energy away with it; in the example above the CO_2 is 'hot' as it leaves the catalyst, due to the high exothermicity of the reaction, and for the safe operation of large-scale reactors this excess energy of reaction must be exchanged, by a variety of heat transfer pathways, with a coolant at the walls of the reactor (Fig. 1.6).

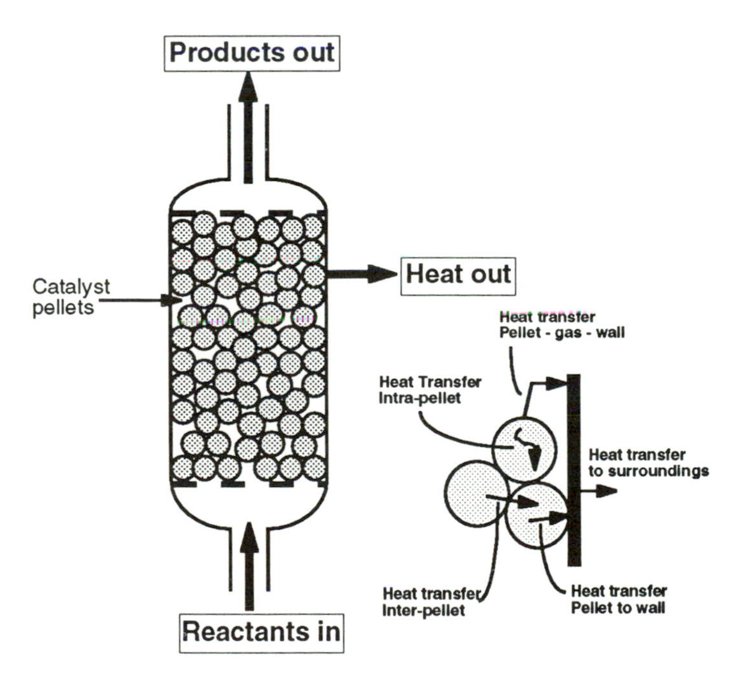

Fig. 1.6 Schematic diagram of heat transfer in an industrial reactor.

It is the availability of bonding sites on the catalyst that favours the catalytic process. This provides a lower energy pathway for molecules to rearrange their bonds in the breaking and reforming that is required for a chemical reaction. This is illustrated in Fig. 1.7. In the gas phase the reaction activation energy ΔE_g is high, largely because energy has to be expended to break bonds, before bond making begins. On the surface, however, the reacting molecules are anchored during the process, always keeping this bonding which stabilises the intermediates in the reaction.

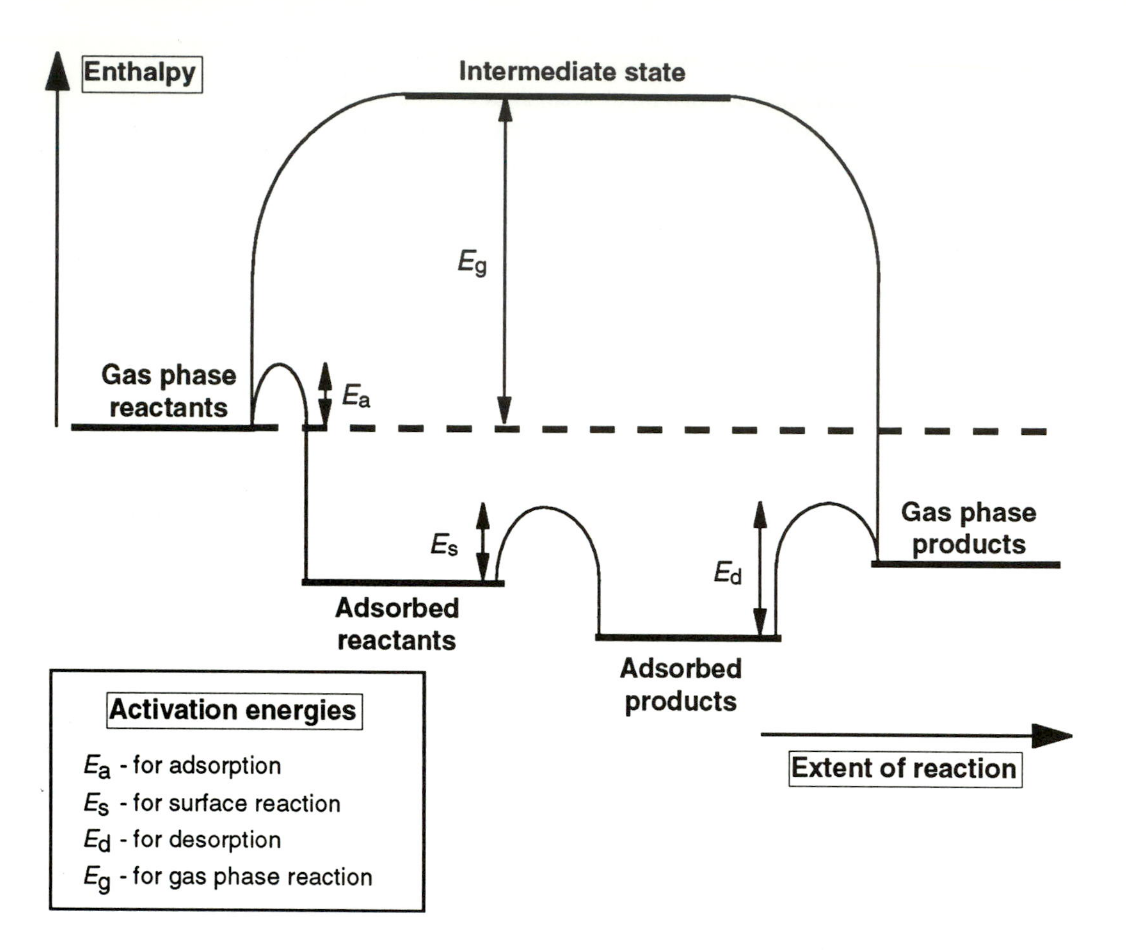

Fig 1.7 The simplified energetics associated with a catalytic reaction.

The result is that the activation barriers are generally lower than for the uncatalysed process and the reaction is kinetically accelerated. A catalyst, then, enhances reaction rates by lowering activation barriers, speeds the approach to equilibrium, but does not affect the equilibrium concentrations.

1.4 Adsorption on surfaces

Figure 1.5 shows adsorption as the first reactive step in the cycle of heterogeneous catalysis. Strong adsorption is called chemisorption and involves real bond breaking or weakening in the reactant and the making of bonds to the surface. In general, as Fig. 1.7 shows, this is a thermodynamically 'downhill' process, at least for transition metals, and is the basis of why catalysis works. The solid surface stabilises intermediates by bonding, which in the gas phase would otherwise have unsatisfied valences and a relatively unstable configuration.

J.E. Lennard-Jones, Trans. Faraday Soc. 28 (1932) 333.

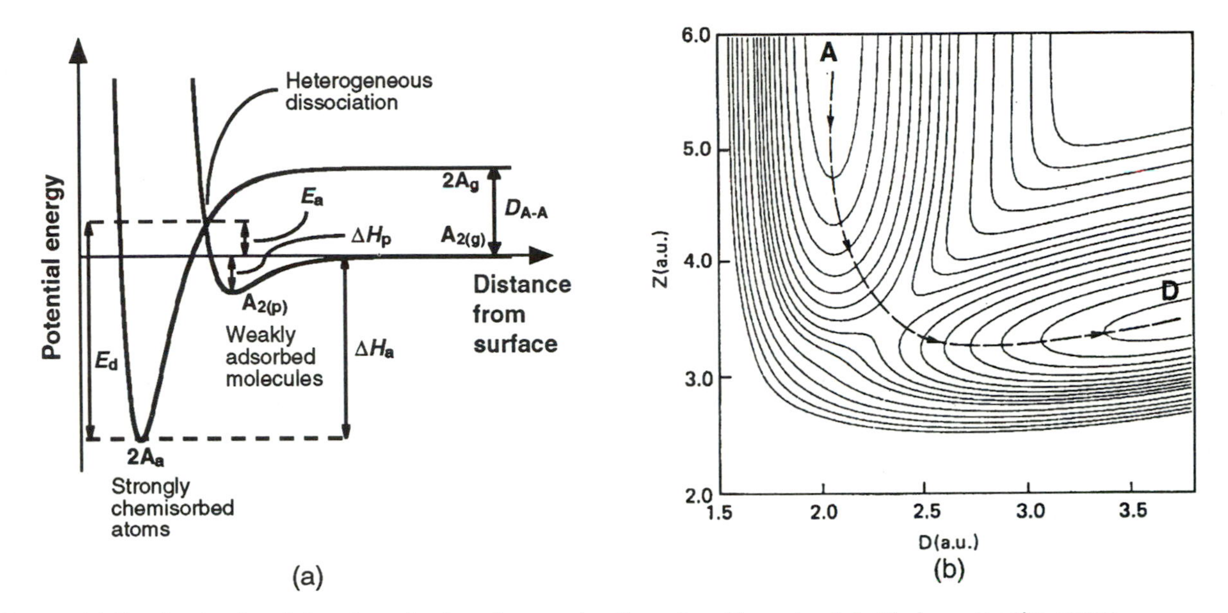

Fig. 1.8 (a) Showing the dissociation of a molecule as it approaches the surface. The molecule first feels an attractive energy minimum which is weak (physisorption, similar to the forces in liquids), but can dissociate from that state. Otherwise it can dissociate direct from the gas phase, if it has energy higher than the activation barrier. For atoms from the gas phase there is only attraction until they are very close to surface atoms, and a deep energy minimum for adsorption. In (b), this curve is represented in an alternative way and shows what happens to the length of the internal bond in a molecule as it approaches a surface for which adsorption is activated. This is a contour map of potential energy and the arrows show the minimum energy path for dissociation between molecular state A and dissociated state D, showing bond extension before dissociation. [Courtesy A. Luntz and J. Harris, J. Vac. Sci. Tech. A10 (1992) 2292.]

The energetics of adsorption can be viewed in a simple way using a Lennard–Jones-type one-dimensional description, as shown in Fig. 1.8a, or, in a somewhat more complicated but more realistic way, in terms of Fig. 1.8b.In Fig. 1.8a two states of adsorption are shown for the case of a molecule that can dissociate. As the molecule approaches the surface it begins to feel an attractive force due to Van der Waals and mainly electrostatic polarisation effects between the molecule and the solid. At a closer approach the molecule begins to experience repulsion (an increase in energy) due to the increasing proximity of the outer electronic orbitals of the solid and the molecule (Pauli exclusion principle). As a result there is a minimum in energy into which the molecule can be accommodated; this state of adsorption is called physisorption and is associated with weak adsorption, very similar to liquefaction, or condensation onto a cold surface. The heat of adsorption in such a state is very low (~20 kJ mol^{-1}), although the exact number depends on the size and complexity of the molecule. The average lifetime of a molecule adsorbed on the surface in this state can be approximated using the Frenkel equation, where τ is the surface lifetime, τ_0 is the lifetime of a surface vibration (~10^{-13} s) and ΔH_a is the heat of adsorption.

$$\tau = \tau_0 \exp^{(\Delta H_a / RT)} \qquad (1.2)$$

The lifetime at three temperatures is given in Table 1.2, for the real example of nitrogen adsorbed at a tungsten metal surface.

If a molecule, such as nitrogen, is dissociated in the gas phase, then it is in a highly energetic state and bonding to any surface lowers its energy, as shown in Fig. 1.8a. In this case its energy is lowered to the extent that it adsorbs into a stable, dissociatively adsorbed state, with a net exothermic reaction, even when taking into account the energy used to dissociate the molecule in the gas phase. Again, as the atom is pushed closer to the surface than its equilibrium position, it experiences repulsion between electrons from the solid and its own non-bonding electrons due to Pauli exclusion. This bonding state is called the chemisorbed state and has a wider range of energies, from ~40 kJ mol^{-1} for weak, molecular chemisorption to ~600 kJ mol^{-1} for the strongest binding atomic species.

Table 1.2 Approximate lifetime of nitrogen in different states on W at 300 K

State	Lifetime (τ)
Physisorbed molecule (ΔH_a ~20 kJ mol^{-1})	3×10^{-10} s
Chemisorbed molecules (ΔH_a ~50 kJ mol^{-1})	5×10^{-5} s
Chemisorbed atoms (ΔH_a ~350 kJ mol^{-1})	10^{40} years

see R. Raval, M. Harrison and D.A. King in "The Chemical Physics of Solid Surfaces", eds Woodruff and King, Volume 3A, p 93 ff (Elsevier, 1990).

It is noticeable that the two curves for chemisorption and physisorption cross in Fig. 1.8a, and this gives the possibility of curve crossing of a molecule as it approaches closely to the surface, from the molecular to the dissociated curve. Fig. 1.8a shows a net activation barrier (E_a) from the gas phase into the dissociated state, though in many cases this barrier may be zero (e.g. oxygen dissociation on most transition metal surfaces). The size of this barrier depends on the solid involved, its surface structure and the nature of the incoming molecule. For instance, for oxygen on most transition metal surfaces this barrier is zero, whereas for Ag it is significant at ~15 kJ mol^{-1}. This then affects the adsorption probability, which can be written in a simplified form as follows:

$$S = \mathrm{k}\exp^{(-E_a/RT)} \tag{1.3}$$

For oxygen on most transition metal surfaces S is close to unity, whereas on Ag at room temperature it is low, at ~10^{-3}, even for the most reactive surface structures. Often, activated dissociation of the latter type occurs directly from the gas phase, without the molecule being accommodated into the physisorbed state, since dissociation from that state is even more difficult (the barrier in that case being equal to the gas phase activation energy plus the heat of physisorption).

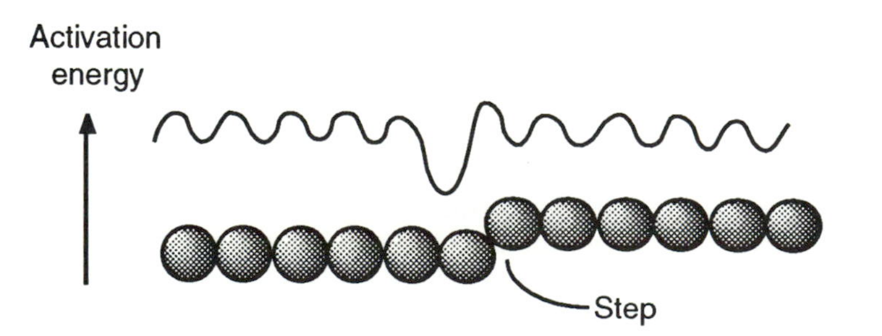

Fig. 1.9 Shows a 1-D plot of the activation barrier to adsorption as a function of distance across a surface with a step. Adsorption is difficult on the flat terraces, but much easier on the step which acts as the 'active site' for adsorption in this case.

The major difference between the simplified diagram of Fig. 1.8a and that of 1.8b is that the former has an axis labelled 'distance from surface'. The problem is that the path taken by the incoming molecule into its final dissociated state may be far from linear and so Fig. 1.8b gives an alternative view in terms of what's happening to the bond distances within the molecule and between the molecule and the surface. For the case, shown, there is a barrier to dissociation from the molecular state, but it is not simply traversed by impacting with the surface with sufficient energy. In this case, if the molecular bond is significantly lengthened before dissociation, then dissociation is easier; a variety of factors can affect the efficiency of this process, including molecular orientation upon approach (rotation), vibrational state and the structure of the surface. Furthermore, as shown in Fig. 1.9, the surface can be considered (at least for highly activated adsorption) to have only certain 'active' areas, where the dissociation barrier is lower than in other parts of the surface. Then a great many collisions with the surface will be ineffective and the adsorption will be inefficient.

2 The effect of surface structure on reactivity

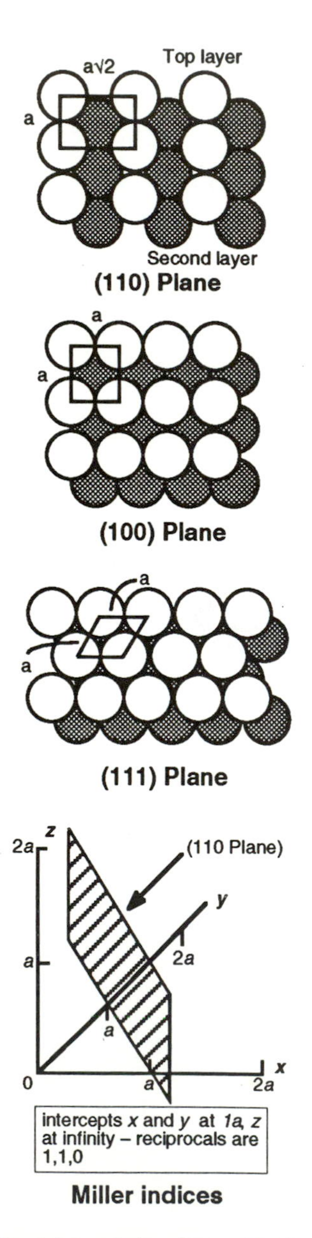

Fig. 2.1 Low index planes of the fcc crystal and Miller index notation.

2.1 Surface structure

Since the nature of the surface involved in heterogeneous catalysis is crucial to performance it is important to consider the structure as a starting point for the discussion of catalytic properties. The surface rests on the bulk and so is related to the bulk structure. In this short text we cannot describe all the types of structure that can be formed, but we will exemplify them by consideration of one class, that is the face-centred cubic (fcc) class. Figure 2.1 shows slices through this crystal which produces the major, so-called low index faces, the (100), (110) and (111) planes. The numbers are Miller indices and their derivation is shown in the margin and is the reciprocal of the intersections of these planes through the major crystallographic axes.

It is evident that the coordination of the atoms in each of these surface planes is different. Generally speaking, surfaces with lower coordination surface atoms have the highest surface free energy, the highest reactivity for adsorption and the strongest binding for the adsorbate (high adsorption heat). The (100) atom has 4 nearest neighbours in the surface, (110) has only 2, while the (111) is in a close-packed structure with 6 nearest neighbours. This has a significant effect on reactivity – the open surfaces generally being the most reactive and the close-packed the least reactive, as described with examples in Section 2.4 below. The surface unit cells are indicated in the figures. These are the three low index faces, but many other orientations are possible. Most importantly, surfaces consisting of mostly well-defined faces of this kind can have other types of morphological features, as shown in Fig. 2.2. Generally, metal crystals cannot be cut to exactly the surface required and so have steps relating to the angle of the cut (usually ~0.5° off the indicated plane). Here there are a range of sites, depending upon the number of nearest neighbours. Those that are most coordinated in the surface are the terrace atoms, next are those at steps, next are those at kink sites within the steps, then double kinks and, finally, isolated adatoms (adsorbed atoms) on terraces. The distribution of these kinds of atoms is a strong function of temperature, but those with maximum coordination in the terrace are the most stable while adatoms are the least stable. In turn, the former will diffuse to new sites slowest, while the latter will diffuse fastest.

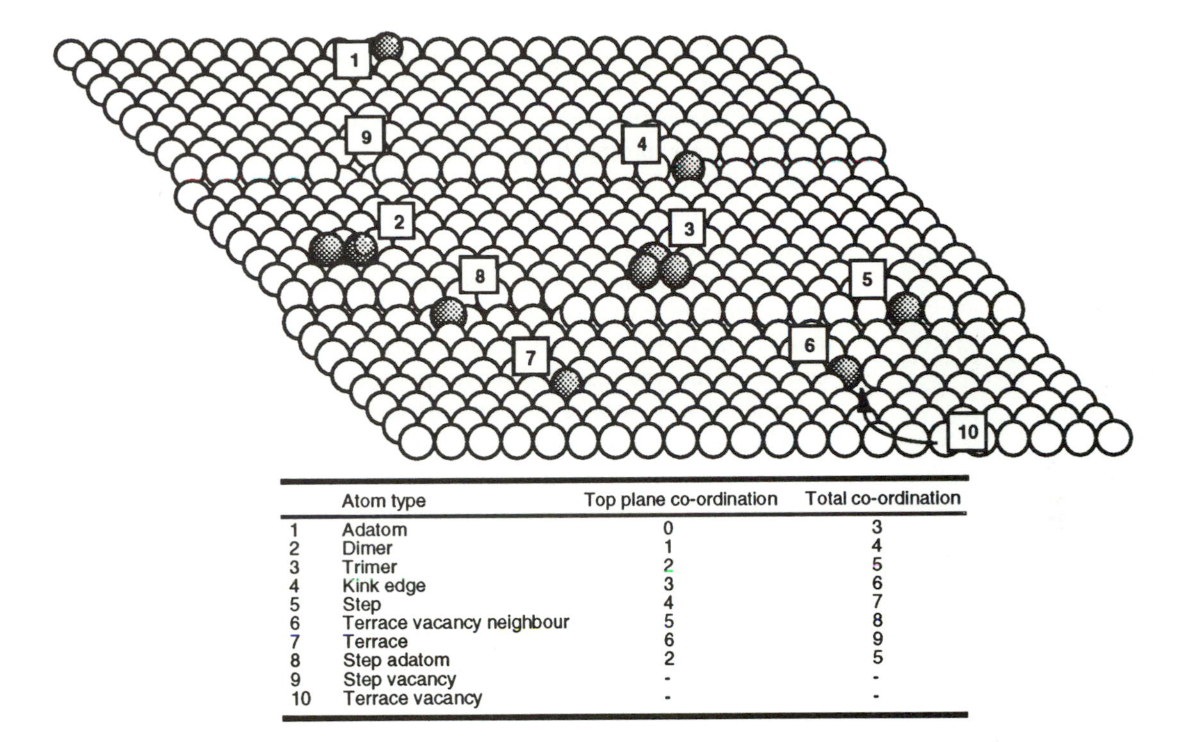

	Atom type	Top plane co-ordination	Total co-ordination
1	Adatom	0	3
2	Dimer	1	4
3	Trimer	2	5
4	Kink edge	3	6
5	Step	4	7
6	Terrace vacancy neighbour	5	8
7	Terrace	6	9
8	Step adatom	2	5
9	Step vacancy	-	-
10	Terrace vacancy	-	-

Fig. 2.2 Crystal surface structure, based on the fcc (111) plane with (100) steps, showing different types of atomic environment.

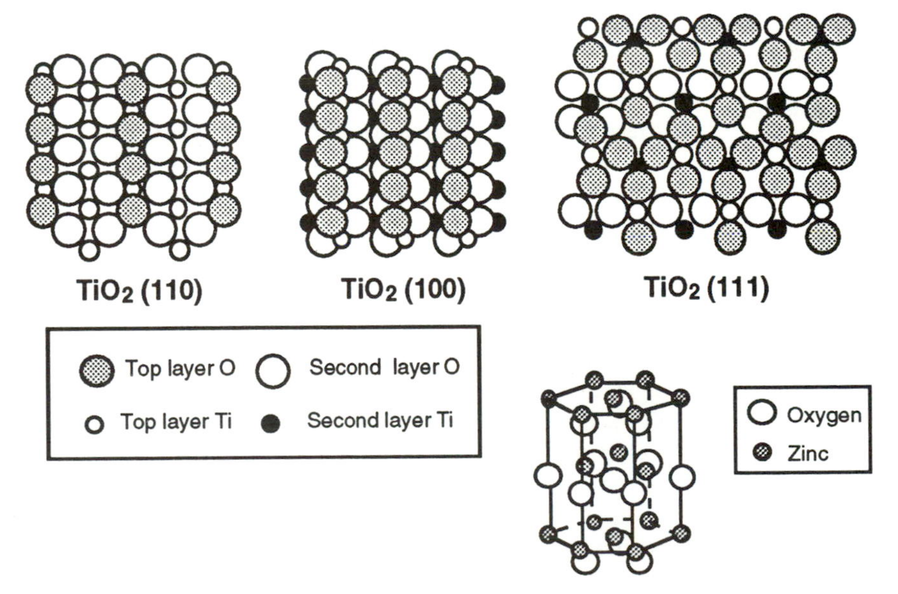

Fig. 2.3 Surfaces of oxide crystals. The low index planes of rutile TiO_2 are shown, together with the bulk structure of ZnO, showing polar and prism faces exposed.

The structures of compound materials are generally more complex, and are too diverse to describe here, but two examples are given, for the low index faces of TiO_2 and ZnO (Fig. 2.3). These faces generally have at least two kinds of termination, some exposing only the anion or cation in the surface plane, some exposing both in the ideal terminations. The so-called polar (0001), $(000\bar{1})$ faces of ZnO, for instance, terminate ideally with only Zn or O atoms in the outermost plane.

The surface of a small metal catalyst particle will often be more complex, and more heterogeneous than this simple picture, comprising a surface with many types of crystallography exposed (more like that shown in Fig. 1.3): this is considered in more detail in Section 2.3 below. However, single crystals of the type described above are widely used in surface studies, in order to examine how surface structure affects reactivity. Using a material with only one type of surface coordination also helps with the understanding of reactivity by eliminating one of the variables involved, that is, the heterogeneity of catalyst surfaces. Great care is taken in preparing such single crystals. They are grown very slowly, usually from a very high purity melt of the metal, as a boule (a cylindrical rod). This is then cut to the appropriate orientation, which is checked by Laue back diffraction of X-rays for accuracy of orientation. The surface is then polished mechanically and/or electrochemically, the former with abrasive pastes down to a sub-micron grade. This process results in a mirror finish to a crystal, as shown in Fig. 2.4. The use of such well-defined surfaces has tremendously expanded our understanding of heterogeneous catalysis in the last 20 years. Single crystals of most metals are readily available, whereas high purity oxide crystals represent much more of a problem, and only very few types are produced.

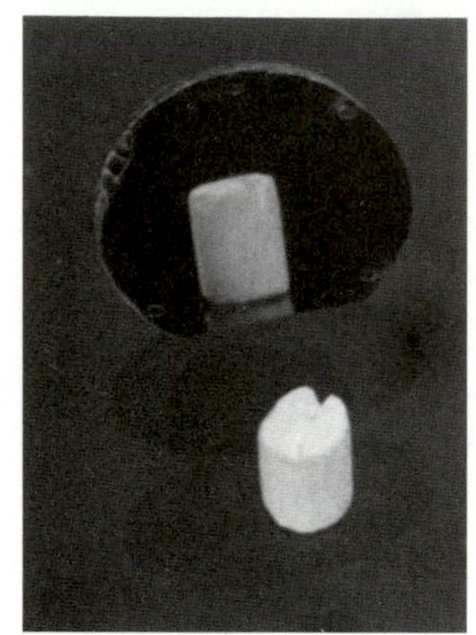

Fig. 2.4 Showing the mirror-like finish of a single crystal surface, reflecting a catalyst pellet.

In the experimental situation it is essential to be able to check the surface crystallography and this is routinely done by back diffraction of low energy electrons (low energy electron diffraction, LEED) as shown in Fig. 2.5. The surface structure is altered by adsorption of gases and LEED can be used to analyse the structure of the adsorbate on the surface. However, nowadays it is becoming common to use STM (scanning tunnelling microscopy) to determine adsorbate structures directly (Fig. 2.6).

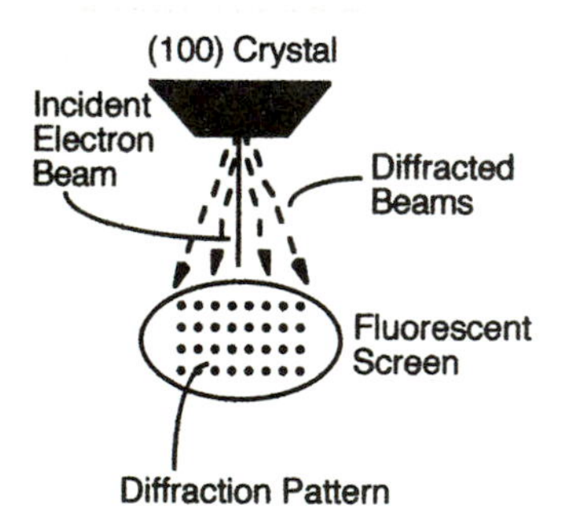

Fig. 2.5 Low energy electron diffraction.

The composition and cleanliness of the surface is determined using surface-specific techniques, again using electrons which have a very short escape depth from the surface region; only those originating from near the surface can escape and be detected. The two major techniques for such analysis are Auger electron spectroscopy (AES) and X-ray photoelectron spectroscopy (XPS). In both, the energy of emitted electrons is analysed, after excitation by electrons or X-ray photons; these energies are element specific and can be used to identify the elements present in the surface region. This is because the energy of emitted electrons is in the range of short penetration depth through a solid. Thus the electrons selected come only from the near-surface region (Fig. 2.7).

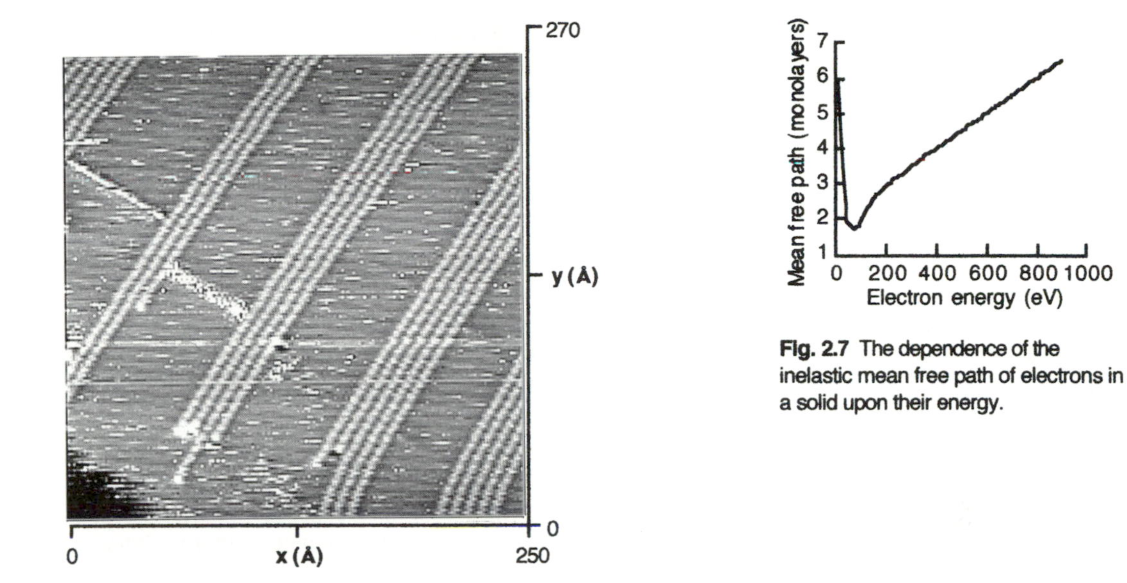

Fig. 2.6 STM showing islands of oxygen adsorbed on Cu(110) in the p(2×1) structure. [Courtesy of the Catalysis Research Centre, University of Reading, and Elsevier from A.H. Jones, S. Poulston, R.A. Bennett and M. Bowker, Surf. Sci. 380 (1997) 31.]

Fig. 2.7 The dependence of the inelastic mean free path of electrons in a solid upon their energy.

Two examples of the application of surface analysis to identifying contamination on surfaces and surface composition are given below (Fig. 2.8).

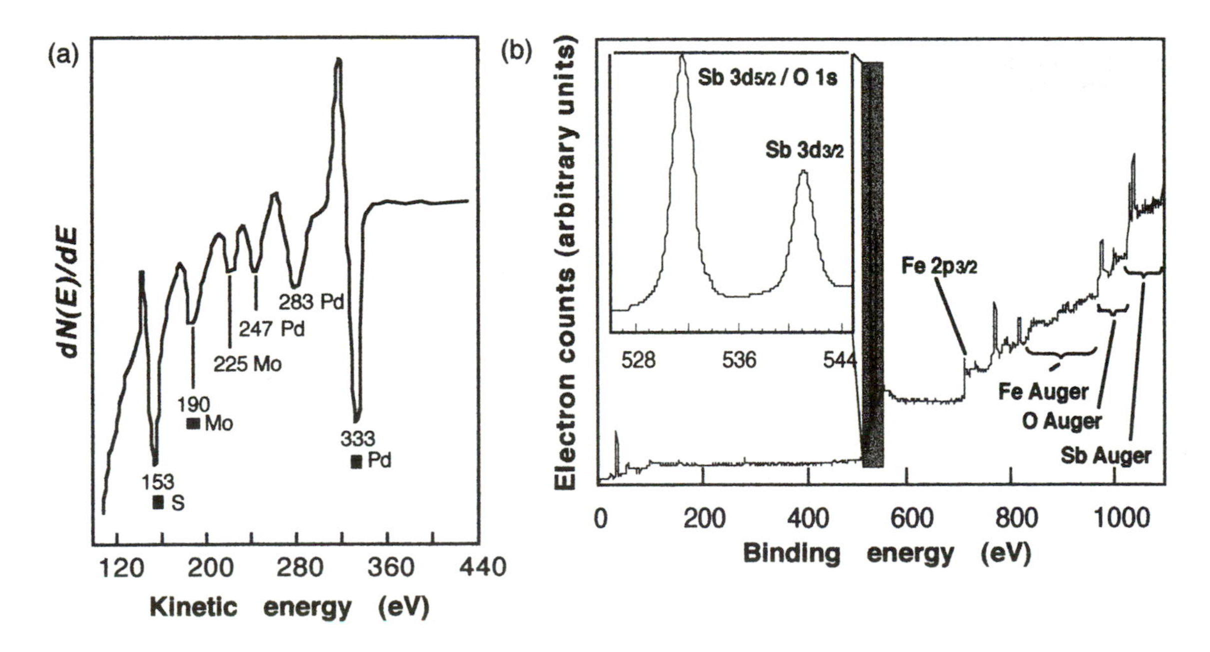

Fig. 2.8 (a) Auger spectrum of a Pd single crystal surface with a high coverage of S contamination; peaks from the Mo support block are also seen. (b) XPS spectrum of a powdered $FeSbO_4$ ammoxidation catalyst, showing preferential segregation of Sb. [Courtesy of the Catalysis Research Centre, University of Reading and Baltzer Science Publishers, from M.A. Allen and M. Bowker, Catal. Lett. 33 (1995) 269.]

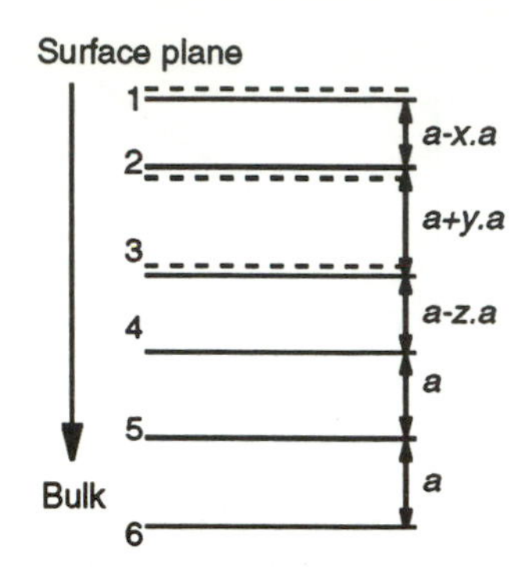

Fig. 2.9 Relaxation of atomic planes in the surface region. Solid lines are real plane positions, dashed are as expected from a continuation of the bulk lattice parameter.

First, a single crystal of Pd contains a low level of bulk sulphur when new, but this material segregates to the surface upon heating; cycles of heating and bombarding the surface with high-energy Ar ions knocks these atoms off the surface and results in a clean interface. The Auger spectrum shows well-defined and separated peaks for the Pd metal and for S which help to identify when the surface is representative of clean Pd. The second example shows the application of XPS to the analysis of an $FeSbO_4$ catalyst used for the oxidation and ammoxidation of propene. It is obvious from the spectra that the component elements are present and that the surface is not of the bulk stoichiometry – it is highly enriched in the Sb component. In fact, it is likely that the surface is almost exclusively Sb and O, since most of the Fe signal comes from material below the surface layer (the inelastic mean free path is several monolayers at this electron energy).

2.2 Surface restructuring

Because the surface is a region of high energy, thermodynamics dictates that there will be a tendency to minimise this energy. In nature this occurs in several ways.

Surface relaxation

All surfaces show a changed lattice parameter in the top layer due to the lack of neighbouring solid atoms on one side of the interface. In general, the surface atoms are pulled closer to the bulk, which in effect increases the coordination of the surface atoms a little, by increased electron overlap with the layers below: this is illustrated in Fig. 2.9. The extent of this relaxation depends on the crystal plane and on the material, but generally ranges from ~2 to 10% of the bulk lattice parameter.

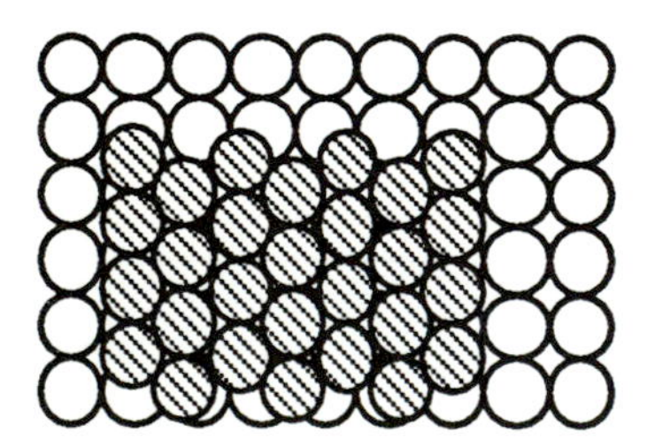

Fig. 2.10 Reconsruction of the (100) Pt surface to form a near-close-packed layer.

Surface reconstruction

Here there is gross reordering of the surface atoms, often resulting in a changed number of atoms in the topmost layer. The most well-known example of this is the reconstruction of Pt(100), which forms a (5×1) surface – the numbers show that the surface unit cell is very large with respect to the underlying bulk (100) structure (Fig. 2.10). This reconstruction was first interpreted by LEED, which shows a large number of spots, many more than expected from the normal termination (Fig. 2.5). A commonly observed reconstruction is the missing row reconstruction of a (110) surface, where alternate atomic rows on the surface are removed. Again, here the surfaces are trying to maximise surface coordination; the (5×1) Pt(100) reconstruction produces a near-close-packed structure, with a surface coordination of 6 rather than the coordination number of 4 for the unreconstructed surface.

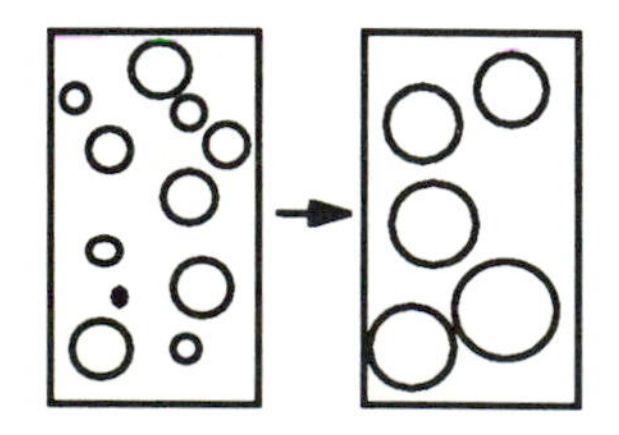

Fig. 2.11 Showing loss of surface area by particle sintering.

Sintering

If a catalyst is produced which consists of separated metal particles, the total surface energy can be minimised by merging of these units to form fewer bigger ones (see Fig. 2.11 and 1.2). This results in lower surface area and higher average surface coordination. This effect is one of the major causes of loss of efficiency of a catalyst that generally occurs during its lifetime in an industrial plant.

Adsorption

Another way of reducing surface free energy is by adsorption of molecules at the surface, resulting in surface compound formation and the satisfaction of free surface valencies. This is most significant in relation to catalysis because it is the first step in the reaction and is the driving force for catalysis. Adsorption can often be accompanied by reconstruction of the metal atoms in the surface. For instance, in the case of Cu(110) the adsorption of oxygen results in the formation of chains of Cu–O units on the surface (Fig. 2.12). The Cu which forms this 'added-row' structure is thought to come from Cu atoms removed from steps. Counter-intuitively reconstructed clean surfaces can form the bulk termination when an adsorbate impinges on the surface. Thus, small amounts of oxygen or carbon monoxide convert the Pt(100) (1×5) surface back to the (1×1). All these effects are driven by the minimised energy situation related to the packing density of the metal and of the adsorbate. In the case of a supported catalyst surface, such effects are likely to be more severe, since, at least for small particles, the coordination is low. Thus, for Rh catalysts, exposure to CO can pull off Rh atoms to form Rh *gem*-dicarbonyl molecules, $Rh(CO)_2$, which are bonded to the support as separate entities.

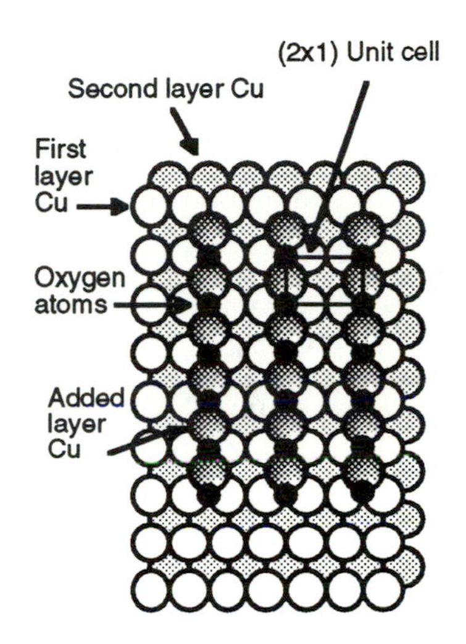

Fig. 2.12 Showing the 'added–row' p(2×1) structure of oxygen atoms on Cu(110).

2.3 The surface structure of catalysts

Using the face-centred cubic structure as the basis, we can build a catalyst atom by atom as shown in Fig. 2.13. This illustrates qualitatively the change in average coordination in the surface as the particle increases in size. The surface energy is related to the number of unsaturated bonds in the surface atom and so as the average coordination goes up with particle size, so unsaturation goes down and surface energy diminishes. There are, however, still individual atoms of low coordination, even on the bigger particles.

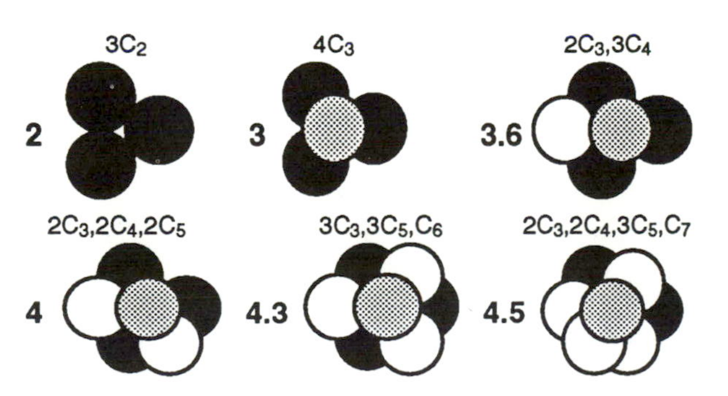

Fig. 2.13 Schematic diagram of the build-up of a small catalyst particle atom by atom, based on a fcc (111) plane. The numbers above the figure refer to coordination e.g. $3C_2$ means 3 atoms coordinated to two others, while the bold figure refers to average coordination of atoms in the particle.

The coordination shown in Fig. 2.13 can be compared with that in Fig. 2.1 and it shows that low-index single crystals may be poor models for the reactivity of small metal particles. However, crystals can be prepared with a high proportion of low coordination atoms (kinks, for instance, see Fig. 2.2). Furthermore, many catalysts have relatively low dispersion metal phases with particle sizes which may expose mainly (111)-type planes. Work on the methanation reaction ($CO + 3H_2 \rightarrow CH_4 + H_2O$) shows very close agreement for the results of single crystal experiments, compared with supported particulate Ni catalysts (Fig. 2.14).

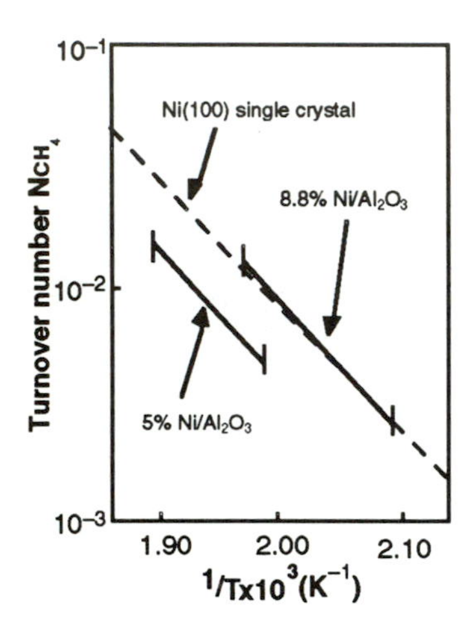

Fig. 2.14 A comparison of rates of CO hydrogenation to methane over a single crystal and over supported catalysts, showing similar activation energies. [Courtesy of Prof. D.W. Goodman and Elsevier; from R.D. Kelley and D.W. Goodman, in The Chemical Physics of Solid Surfaces and Heterogeneous Catalysis, Vol 4 (1982) 36.]

In catalysis the number of atoms at the surface of a particle, compared with the total atoms, is often called the dispersion, and the formula for dispersion is given below. The relationship between surface atoms and particle diameter is given in Fig. 2.15 for any hemispherical body made up of spherical sub-units which are ordered (the cross-section of the particle is assumed to be connected to a support). This can be more generally represented by the following equation for the surface area (*SA*)

$$SA\ (m^2) = \frac{3V}{r} = \frac{3W}{\rho r} \qquad (2.1)$$

where V is the total volume of material in the catalyst, r is the average particle radius, ρ is the density of the material and W is the weight of the sample. This relationship derives from the numbers of atoms in the surface and bulk phases. For such a hemispherical particle, the exposed area is

$$SA = 2\pi r^2 \qquad (2.2)$$

and the volume of the particle is

$$V = \frac{2\pi r^3}{3} \qquad (2.3)$$

If, instead of r, we use N for the number of atoms in the radius of a particle, then the number of atoms at the surface, N_S is given by

$$N_S = 2\pi N^2 \qquad (2.4)$$

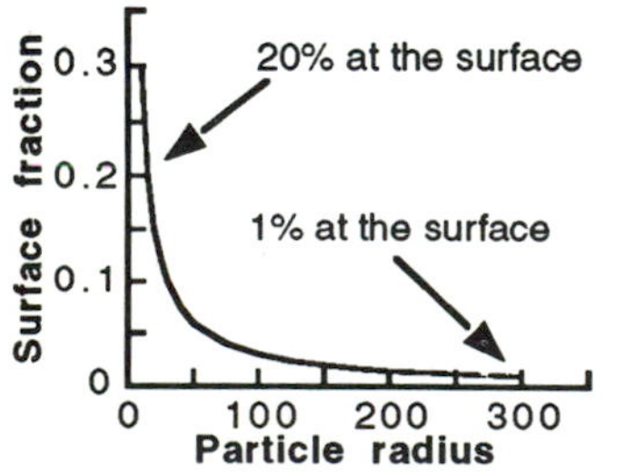

and the total number of atoms in the particle by

$$N_T = \tfrac{2}{3}\pi N^3 \qquad (2.5)$$

and so dispersion can be described as

$$D(\%) = \frac{N_S}{N_T} \times 100 = \frac{3}{N} \times 100 \qquad (2.6)$$

Fig. 2.15 The relationship between fraction of atoms at the surface and the particle radius.

This spherical approximation does not apply accurately to the smallest particles, but is quite a good approximation for larger ones. If the shape of the particle is different (for instance, cubic) then similar simple expressions apply, but with a different multiplier (5 for a cube with one face on a support material).

The measurement of the shape and surface of small particles can be a very difficult problem, and this is dealt with in more detail in Chapter 6. However, for larger particles some idea of surface morphology can be obtained from imaging techniques such as SEM (see Fig. 1.2), or high-resolution TEM (transmission electron spectroscopy) (Fig. 2.16a) or even by using the new technique of scanning tunnelling microscopy (STM) (Fig. 2.16b) on model catalysts.

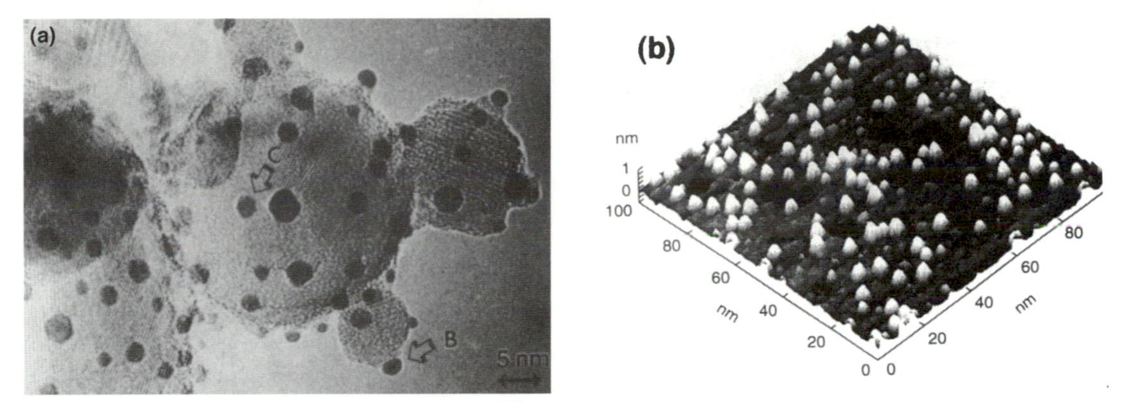

Fig. 2.16 (a) Showing a transmission electron micrograph (TEM) image of Au supported on TiO_2 microspheres. [Courtesy of American Chemical Society, from G. Fuchs, D. Neimien and H. Poppa, Langmuir 7 (1991) 2856.]; (b) Showing an STM image of small particles of Ni on alumina. [Courtesy of Prof. D.W. Goodman and Royal Society of Chemistry, from C. Xu, X. Lai and D.W. Goodman, J. Chem. Soc. Faraday Discuss. 105 (1996) 256.]

An important factor affecting some oxidic and all pure metal catalysts is their relative lack of thermal stability, which generally results in these materials having a very low area due to sintering. Thus, in almost all cases the active phase is 'supported' on an inactive, but thermally strong, support phase which is ceramic and refractory in nature. The most common such supports are alumina, silica and carbon. All of these can be obtained in very high surface area forms (up to c. 200 m^2 g^{-1}) and they maintain their areas even in very harsh temperatures and environmental conditions. Hence, in order to produce high area active phases of metals and oxides they are prepared in a form that is bonded to the support phase (see, for instance, Fig. 2.16, Fig. 2.20 below and Fig. 1.2). This keeps the active particles apart and helps maintain their activity over a long period of time. Because of the particulate nature of these materials they are highly porous and high area samples generally have a range of pore sizes, from less than a nanometre or so in diameter (micropores), through to mesopores (~10–100 nm), to macropores (bigger than 100 nm). The distribution of these pore sizes can be an important factor in making a catalyst that has a very good performance.

2.4 Structure dependence and independence of catalytic reactions

Some catalytic reactions show little dependence on the nature of the surface structure involved (see Fig. 2.14 for instance), while others are strongly dependent on it. The best evidence for the influence of structure on a reaction comes from studies using model catalysts, namely single crystals. A particularly useful example of this is hydrocarbon reactions with Pt surfaces carried out by Somorjai's group in California. Fig. 2.17 shows the results of measurements of both the hydrogenation and hydrogenolysis of hexane, showing that the latter is strongly structure dependent, in contrast to the former.

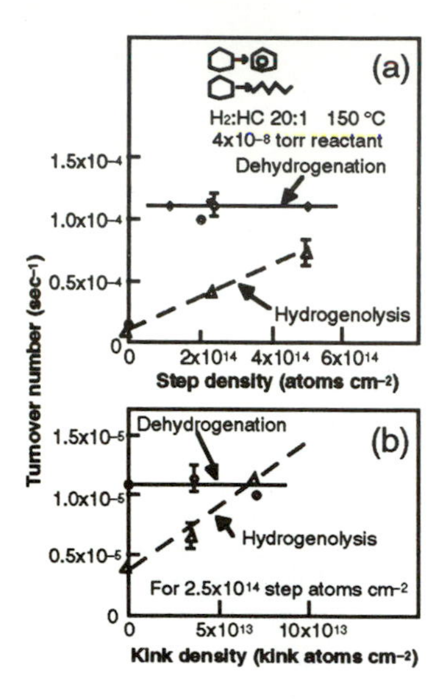

Fig. 2.17 This shows structure -dependent (hydrogenolysis) and -independent (hydrogenation) reactions on Pt crystal surfaces with different densities of defect sites [(a) steps and (b) kinks]. [Courtesy of Prof. G.A. Somorjai and Academic Press, from D.W. Blakely and G.A. Somorjai, J. Catal. 42 (1976) 181.]

Another example concerns the Haber process of ammonia synthesis. Experiments carried out on single crystal Fe surfaces under low pressure conditions shows a marked variation of the dissociative adsorption probability of nitrogen (Fig. 2.18), with the most open low-index face, the (111) for a body-centred cubic crystal, having the highest probability, and the close-packed (110) having the lowest. This again reflects differences in surface free energy and coordination. However, particular types of sites are also supposed to be important for the conversion of molecularly adsorbed nitrogen into atoms, which is thought to be the rate-determining step for this reaction. Experiments carried out in a different laboratory showed that these results were in agreement with the trends of ammonia synthesis activity under conditions of high pressure and temperature near those used industrially (Fig. 2.18). Thus, structure has a controlling influence in this case and modelling calculations confirm that the association of molecular nitrogen is rate-determining.

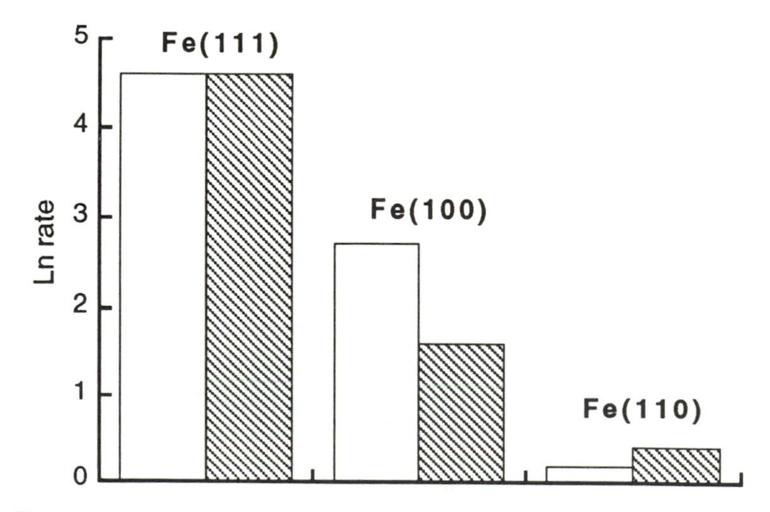

Fig. 2.18 Showing the dissociative adsorption probability of nitrogen on Fe single crystals, (filled bars) which is of a similar sequence to that for ammonia synthesis under near industrial conditions (empty bars). The data have been normalised to the same rates in each case for the (111) plane to show the trend with crystal surface. (Data courtesy of Prof. G. Ertl, Fritz–Haber Institut, Berlin and Prof. G.A. Somorjai, Dept of Chemistry, University of California, Berkeley.)

In general, for dissociative adsorption, the dissociation probability is strongly dependent on the structure of the surface and there is often a more strong dependence on crystal structure than on the particular metal involved (see Table 2.1). A general picture of this is shown in Fig. 2.19 where, open planes [such as (110)] have a high sticking probability into the dissociated state, but close-packed planes have lower values, sometimes orders of magnitude lower. The main reasons for this are the lower work functions on open planes, allowing easier transfer of electrons into anti-bonding LUMOs on incoming molecules, and also the easier approach of the molecules to the electron-rich surface plane. Promotion of a surface with an alkali has a similar effect on a close–packed plane to that when an open surface is used – it induces a rougher surface with a lower work function.

Table 2.1 O_2 dissociation probabilities on various fcc surfaces

Metal	Sticking coefficient	
	(110)	(111)
Rh	0.6	0.5
Pt	0.5	0.05
Cu	0.2	10^{-3}
Ag	10^{-3}	10^{-5}

LUMO Lowest Unoccupied Molecular Orbital

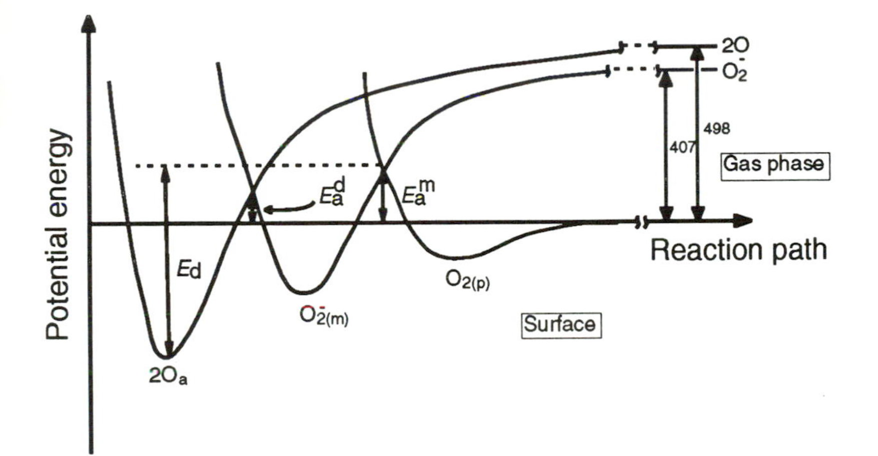

Fig 2.19 Potential energy profile for adsorption of oxygen onto a metal surface showing the possibility of two barriers to adsorption from the gas phase. These barriers are between physisorbed [$O_{2(p)}$] and chemisorbed molecules (O_2^-), and between the latter and the dissociated state. These three states are successively more strongly bound to the surface. High energy gas phase atoms are stabilised by the surface.

These effects are often expressed in catalysis as 'demanding' and 'undemanding' reactions. This feature of the catalysis is usually determined by examining how the rate of reaction per site varies with changing the particle size of the active phase of the catalyst. In general, for reactions where the surface reaction is rate limiting, structure sensitivity is to be expected.

Many catalysts are metallic in nature and are often composed of highly dispersed, small diameter, precious metal particles. There are many additional effects of changing the structure of such a material which, in turn, affect its catalytic activity, as illustrated in Fig. 2.20. When a particle is very small ($\leq$3 nm) it becomes less like a metal because of the splitting of the conduction band into discrete energy levels and this can change the surface reactivity and activation energies for all the steps involved. Furthermore, there can be charge transfer between the catalyst particle and the support material, depending strongly on the nature of the support, on the nature of the metal and on its structure/size. Edge effects can also be very important for small particles where the edge atoms in contact with both the support and the gas–solid interface can be a significant fraction of the atoms involved.

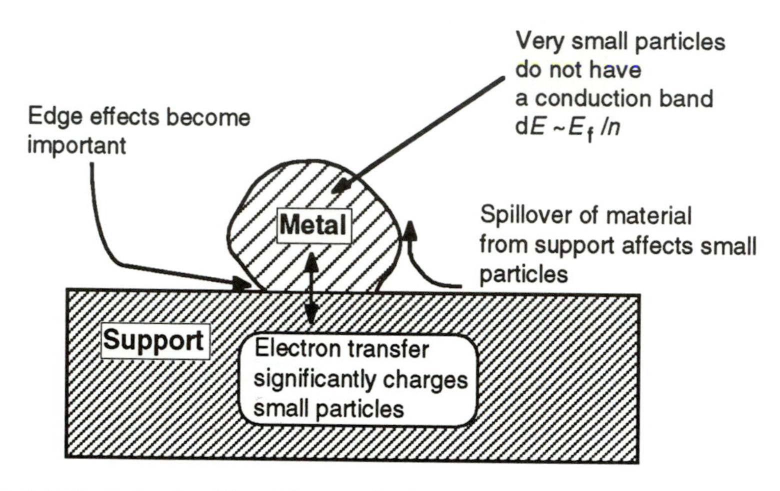

Fig. 2.20 Illustrating the different factors affecting the properties of a small metal particle adsorbed on a support material.

Finally, in recent times an effect known as SMSI (strong metal support interaction) has been discovered and this can occur when metals are supported on partially reducible oxides such as TiO_2. In this case, reduced support compounds can diffuse onto the metal, and this then strongly affects the catalytic performance.

3 Changing the reactivity of catalytic surfaces

3.1 General

If we consider any particular reaction then, in principle, it is possible to manipulate the properties of a catalyst for that reaction by any process that alters the properties of its surface. In thermodynamic terms we may increase or reduce the surface free energy, in microscopic terms we may alter the nature of the individual sites at the surface which are responsible for giving the desired products. Table 3.1 lists the ways in which a particular catalyst can be manipulated, and it is in the engineering of the surface in these kinds of ways that the industrial catalytic chemist should be adept. The practical application of catalytic know-how is to maximise the time–yield integral shown in Fig. 3.1.

Table 3.1 Ways of affecting catalytic performance

Catalyst treatment	Probable effects	Side effects	Achieved by
Sintering	Loss of activity	Change in selectivity due to changed site type ratio; improved longevity	Thermal/oxidation treatment
Dispersion	Increased activity	Change in selectivity due to changed site type ratio; decreased longevity	More controlled preparation, *in situ* gas treatments (e.g. dispersion of Rh in CO)
Promotion	Increased activity	Can change selectivity, longevity	Addition of alkali salts (e.g. KNO_3) during or post-preparation
Poisoning	Decreased activity	Can improve selectivity (e.g. Cl in ethylene epoxidation)	Addition of electronegative elements in preparation or during use
Change support	Increased or decreased activity/selectivity depending on dispersion and charge transfer	Can be secondary poisoning or promotion depending on reducibility of the support	In preparation use of different oxidic materials or precursors
Alter pore size	Can alter activity if diffusion limited	Some shape selectivity on reactants/products for microporous materials. Can effect secondary chemistry in oxidation reactants	Use of supports of different structure, or altered preparation method (e.g. ageing or calcination)

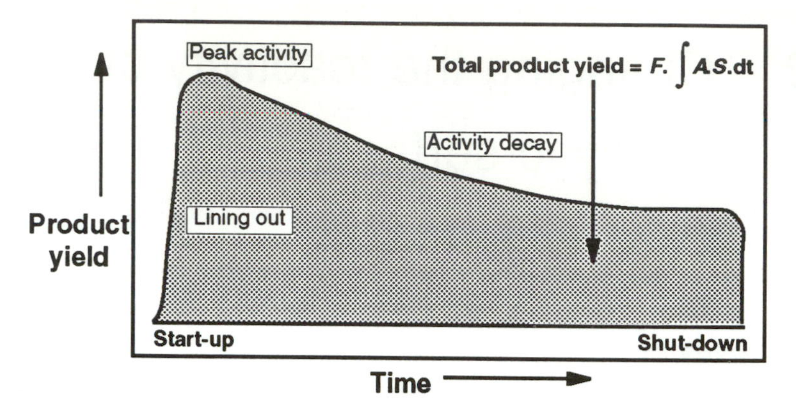

Fig. 3.1 The time dependence of catalyst activity to the desired product in a commercial plant operation. The total yield over the life of the plant is dictated by the flow rate (F) multiplied by the time integral of total conversion activity (A) and selectivity (S) to the desired product. For definitions see Chapter 5.

Typically, a catalyst goes through a 'lining out' period during which the performance is unstable as the surface of the catalyst changes, by sintering, segregation and by achieving steady state in surface concentrations of reactants and products. The performance then usually reaches a maximum and thereafter declines. At some point it becomes no longer economical to continue running the catalyst at lower activity and so the plant is shut down and recharged with new catalyst. The time-scales of these changes can be very variable depending on the catalyst and process, but can vary from days to years. The industrial scientist maximises this yield by developing varied catalysts and testing them in small-scale reactors. However, the inputs to catalyst design and preparation come from a variety of sources, ranging from fundamental surface science to reactor engineering. With the advent of expert systems and high power computing, catalyst design is close to being a prescriptive discipline.

In what follows we will concentrate on the effect of additives located on the surface of the catalyst upon the way catalysis proceeds. The effects of altering other parameters have already been considered above. For instance, it is generally the case that changing the size of the active particle alters its activity – smaller particles being more reactive due to their higher surface energy. However, at least for very small particles (less than *c.* 3 nm diameter), other factors can also be affected by sintering. For very small particles, charge transfer to/from the support can be significant, as described above, and can affect the catalysis positively (increase desired product) or negatively (decrease desired product) depending on the nature of the reaction. If reactions at the interface between the support and active phase are important, they become decreasingly important as particle size increases. Sometimes, however, it may be necessary to have lower activity sites present, either to prevent self-poisoning with strongly adsorbed material, or to favour more selective synthesis; as the particle size increases, the close-packed surfaces are favoured and these may be important for some reactions.

Altering the support material can have dramatic effects for many processes, for a variety of reasons. These are often most significant when the active phase is highly dispersed. Under those circumstances there is often an electronic interaction with the support, which can result in net charging of the particle; boundary effects may also then be important. Such effects are not pronounced for large particles.

We shall consider the effects of engineering the active phase of the material by placing additives at the surface. In practice, most industrial processes have additives that act as modifiers to the chemistry at the surface.

Table 3.2 Major effects of promoters

Major effects of promoters
Activity enhancement
Selectivity improvement
Increased catalyst lifetime
Neutralisation of acid sites

3.2 Promotion and poisoning

Promoters can be classed as substances which, when added to a catalyst as a minor component, improve one or more of the properties of the material with respect to product formation. These effects are summarized in Table 3.2. In very many cases an alkali metal is the promoter and most often it is potassium. A major reason for this is that the compounds of this material (such as the oxide) have low surface energies compared with the metals, which in turn means that they generally segregate to the surface of the catalyst. Thus, although the alkali may be present at only the 1% level, its concentration at the active interface, the surface, can be much greater than this.

An example of this is in ammonia synthesis catalysis, in which nitrogen and hydrogen are reacted together at high pressure and temperature over an Fe catalyst. Here, potash is added to the material during preparation at the 1% level, yet surface analysis of the material shows that it covers the outermost layers of the catalyst to approximately 10% (discussed further in Chapter 9). Table 3.3 gives a list of some of the large-scale industrial catalytic processes that use promoters.

It was stated above that promoters improve the properties of the catalyst for the reaction, but this can occur in several ways.

(a) Activity enhancement. An alkali can directly increase the rate of synthesis.
(b) Selectivity enhancement. A promoter may change the pathway of a reaction to enhance the rate of formation of the selective product.
(c) Lifetime enhancement. Sometimes a promoter can appear to have little effect on the initial performance of a catalyst, but can keep it operating longer (see Fig. 3.1). This is often due to either a reduction of the sintering rate or reduction in the build-up of an irreversible poison on the surface with time, due to alteration of the surface chemistry.

Table 3.3 Some large-scale industrial processes using promoted catalysts

Process	Catalyst/promoter
Ammonia synthesis, Fischer–Tropsch	Fe/K
Ethylene epoxidation	Ag/K, Cs
Methanation, Steam reforming	Ni/K
Water gas shift	Fe oxide/K
SO_2 oxidation	V oxide/K
Oxychlorination of ethene	Cu chloride/K
Propene ammoxidation	Various mixed metal oxides/various promoters, often K
Butane to maleic anhydride	V phosphate/alkalis

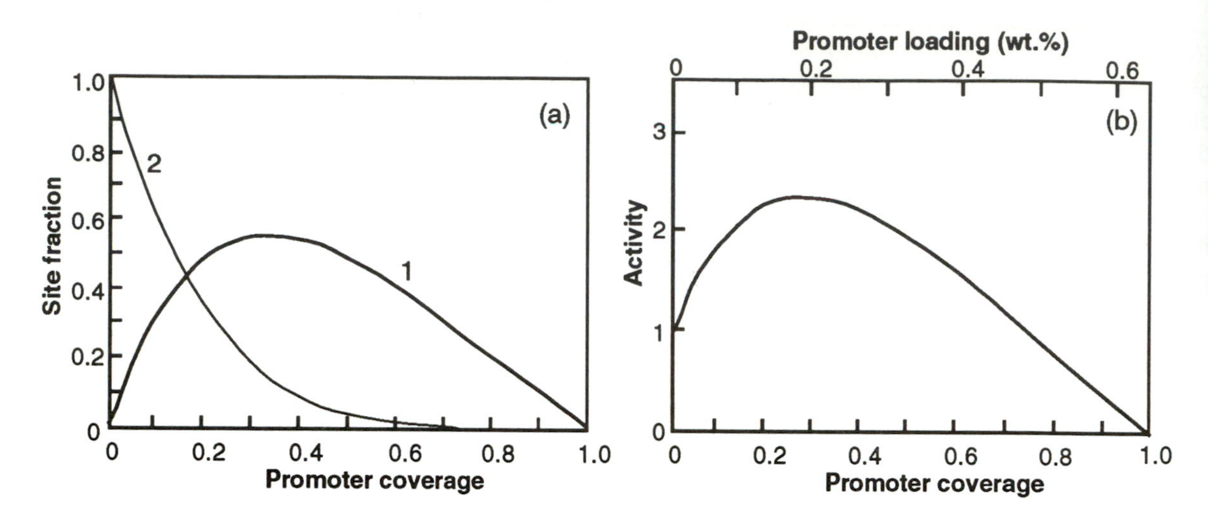

Fig. 3.2 Showing the effect of promoter coverage on (a) the site distribution on a catalyst. The line labelled 1 is the promoted site fraction, while the line labelled 2 represents the unaffected sites for four affected sites next to the promoter. (b) The effect on activity, showing a maximum at low promoter coverages (assuming that promoted sites are 4 times as active as unpromoted ones). The loading at the top of the figure assumes a surface area for the catalyst of 10 $m^2 g^{-1}$, a solid Fe catalyst and a K promoter. [Figure courtesy of Elsevier, M. Bowker, Appl. Catal. 45 (1988) 115.]

Promoters are usually alkali metals, but can vary more widely than that. More generally, they are electropositive elements, often positive ions. An activity promoter operates in the following manner. It activates sites around it, but always acts as a blocker of the site it sits on, since the promoter itself is not active. The effect of this on the site distribution and on the activity of the material is shown in Fig. 3.2.

The model of the working of a promoter has been the subject of much debate over the years, but a consensus is beginning to emerge with the help of theoretions. This can be pictured as in Fig. 3.3. The ionic species generates an electrostatic field is located at the surface, with the positive part of the field outermost. This field in the region of the promoter and interacts with orbitals of incoming molecules in a way that pulls these levels down in energy such that electron exchange between the solid and the reactant can take place more easily. This electron exchange is often the essential requirement of the initiation of the catalytic process. This changes the binding energies of such molecules, strengthening the binding and easing the pathway of dissociation. These effects have been carefully measured for a number of adsorption systems, an example being oxygen dissociation on silver. The addition of a potassium promoter to a silver catalyst increases the dissociation probability by three orders of magnitude from 10^{-6} to 10^{-3}, largely due to a decreased activation barrier to dissociation.

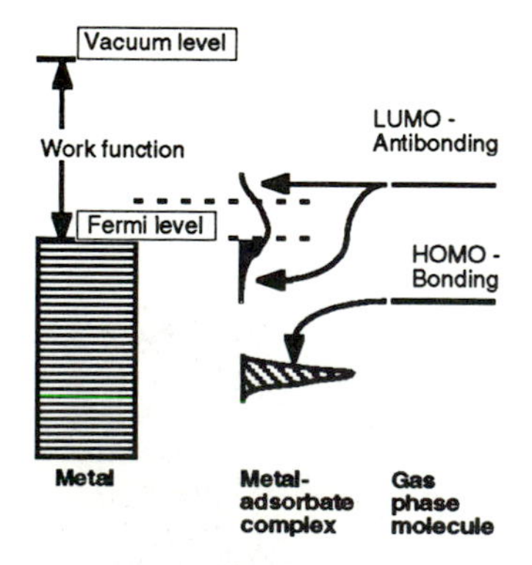

Fig. 3.3 Showing the effect of a promoter on an incoming molecule. A major effect is to lower the energy of LUMOs to allow easier transfer of electrons into them, such orbitals are often antibonding for the molecule and so this effect enhances dissociation.

Poisons tend to be electronegative elements, such as Cl, S, P and C. At the simplest level they block sites on the surface and so reduce the total number of centres of activity. However, they often also induce a field of the opposite polarity to the promoters, which in turn raises molecular energy levels relative to the unpoisoned surface, reduces electron transfer and weakens adsorbate binding. A nice example of this is in the adsorption of CO on nickel where both the total amount that can be adsorbed and the temperature stability is reduced, as evidenced by the desorption experiment in Fig. 3.4 which measures the adsorbed gas evolving as the surface is heated.

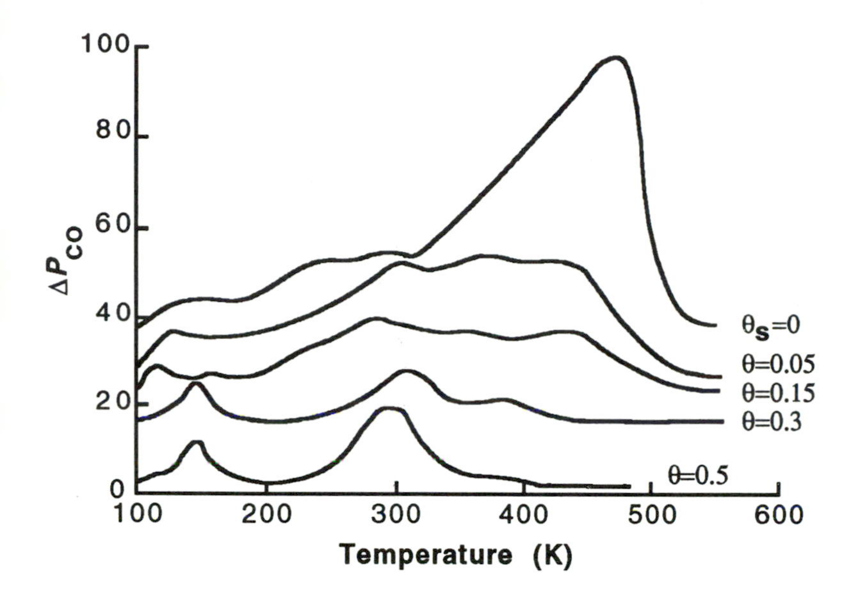

Fig. 3.4 Desorption of CO from a Ni(100) surface, showing the effect of S poisoning. S blocks sites (reduced desorption integral), and reduces the strength of binding (diminished temperature of desorption). [Courtesy of Prof. D.W. Goodman and Elsevier, Appl. Surf. Sci. 19 (1984) 1.]

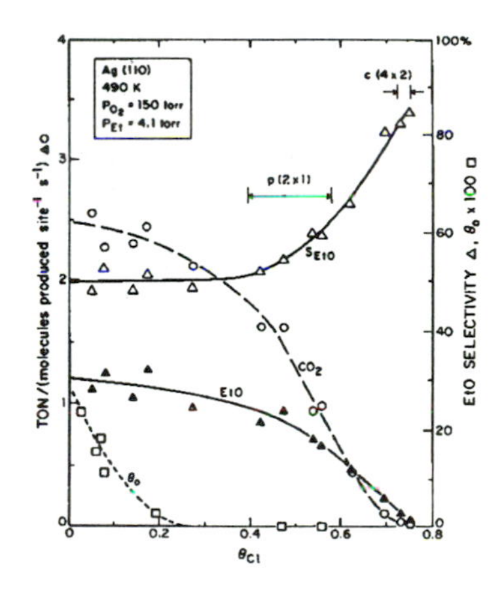

Fig. 3.5 The variation in selectivity to ethylene oxide (EtO) with dosing the Ag(110) surface with different coverages of Cl atoms. Note the poisoning effect on both reactions, but it is more marked for CO_2. [Courtesy of Prof. C.T. Campbell and Elsevier, Appl. Surf. Sci. 19 (1984) 32.]

Poisons can have a positive effect for certain reactions. For instance, in ethylene epoxidation on Ag catalysts, the addition of ppm levels of an organochloride such as ethylene dichloride, leads to the deposition of Cl atoms on the surface and this has a positive effect on the reaction. It improves selectivity from ~50–70%, by preferentially poisoning the non-selective combustion reaction (Eqn 3.2) below.

Selective route,
oxidation $C_2H_4(g) + O(a) \rightarrow C_2H_4O(a) \rightarrow C_2H_4O(g)$ (3.1)

Non–selective route,
combustion $C_2H_4(g) + 6O(a) \rightarrow 2CO_2(g) + 2H_2O(g)$ (3.2)

This is illustrated nicely by single crystal results on Ag(110), which shows this effect at high coverages of Cl on the surface (Fig. 3.5). This is thought to be due to restriction of the transition state during the reaction, or to reduced availability of surface oxygen (Fig 3.6). For the former, the transition state for complete oxidation is bigger than for the selective route, because an intermediate is involved which contains at least one more oxygen atom in the complex than in the selective product, ethylene oxide. Alternatively, Cl blocks sites at the surface where oxygen atoms could be

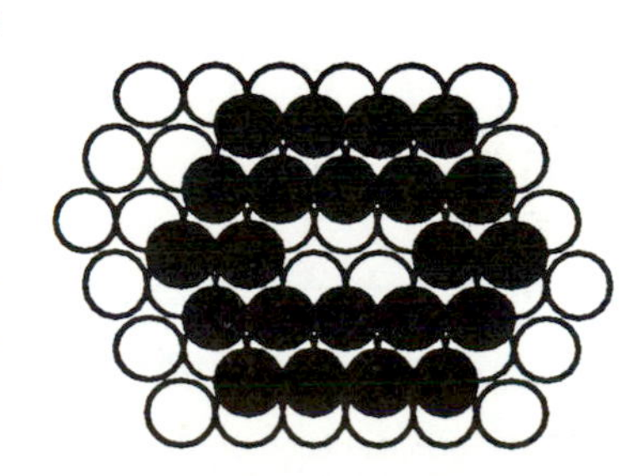

Fig. 3.6 Site isolation during ethylene epoxidation. This is a model of a Ag(111) surface (open circles) partly poisoned by Cl ions (filled circles).

adsorbed and therefore limits oxygen availability in the reaction, slowing up the combustion reaction which Eqn 3.2 shows to be highly oxygen demanding.

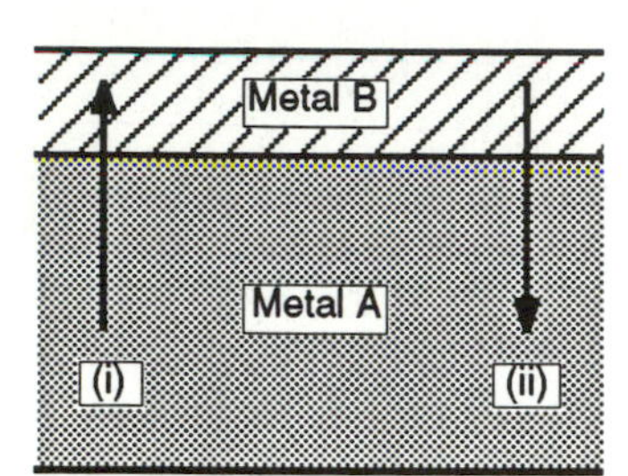

(i) if $d_B >> d_A$

(ii) if $2\Delta H_f^{A\text{-}B} >> \Delta H_f^{AA} + \Delta H_f^{BB}$

Fig. 3.7 Schematically illustrating important effects in alloy formation. d_B and d_A are the diameters of the two metal atoms, while the ΔH_f refers to the heat of formation of the respective bonds in the individual metals and the alloy. (i) Implies segregation of the bigger component to the surface and poor mixing while (ii) implies alloy formation.

3.3 Alloying

The properties of metal or oxide catalysts can be altered by mixing in another metallic or oxide component. Mixed metal alloy catalysts are now widely used in the oil reforming industry. For example, the addition of Re to Pt-based catalysts for naphtha reforming (see Chapter 7) in the early 1970s led to a big improvement in the efficiency of the process. This was largely due to an enhanced lifetime in the reactor, due to reduced carbon poisoning susceptibility. A variety of other alloys can also be used for this process.

This is just one example, in which the effect of the second metal appears to be mainly on reducing poisoning. However, alloying can have a wide variety of effects which are determined both by the distribution of the second metal in the first and by its chemical properties. Where a second component locates largely depends on the relative sizes and bond strengths of the two elements (Fig. 3.7). If there is a large mismatch of atom sizes, then the larger atom tends to be displaced from the bulk to the surface (segregation) and may be totally phase separated from the first. Thus, for Cu and Ag, since the latter is a much bigger atom, it will be pushed out to the surface and indeed these metals do not form good bulk alloys, whereas Ag and Pd, with much closer atomic sizes, do.

If the bond energy of one metal is much less than another (that is, lower cohesive energy, lower heat of vaporisation) it will also tend to segregate preferentially to the surface, since the stronger bonding element will maximise its coordination in the bulk and surface energy is minimised by terminating with the weaker bonding element. For cases where the atomic size and bond energy are similar, there is a delicate balance of population which depends sensitively on the details of these differences of surface structure, temperature and gas phase composition. A nice example of this specificity is the surface structure of a Cu–Pd alloy with 15% Pd in the bulk. In this case the surface terminates with an all Cu outer layer and a 50% mix of Cu and Pd in the second layer.

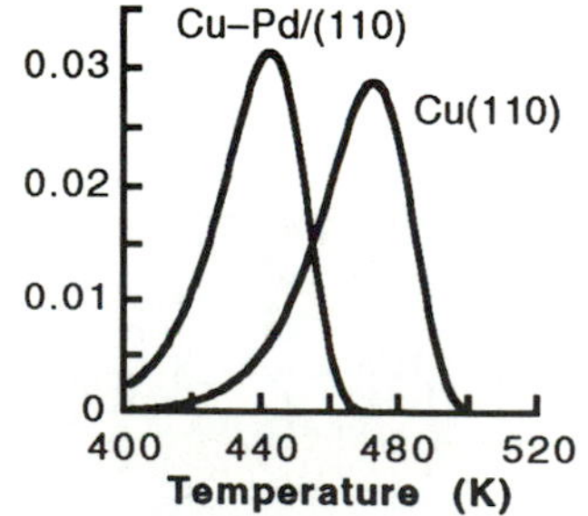

Fig. 3.8 Formate decomposition on Cu(110) and a Cu–Pd(110) alloy, showing the higher activity of the latter, even though its surface consists of only Cu. [Adapted from Newton et al. Surf. Sci. 259 (1991) 45.]

The chemical effects of alloying are severalfold and again depend on the nature of the solvent and solute atoms. In general, if an 'inert' metal like Au is placed on an active metal, such as Pt, the effect is to dilute the active sites, but it can be more marked than this. If a reaction requires a large ensemble of sites to facilitate it, a small coverage of the second metal can have a dramatic effect on the activity for that step, indeed it can act like a poison. On the other hand, the electronic structure of a metal is intimately connected to the catalytic reactivity, and alloying of d-band metals changes this electronic structure. In turn, this affects the stability of surface intermediates and therefore changes the kinetic pathway and selectivity of catalytic reactions. This effect on stability can be seen in Fig. 3.8, which shows that the alloy described above, with only Cu in top layer, is nevertheless more active for formate decomposition to CO_2 and H_2, producing a peak at lower temperature in temperature-programmed desorption (TPD) than for Cu(110).

Mixing metals in oxides also dramatically affects reactivity. When Sb_2O_4 is used for the oxidation of propene, it has very low activity, but is selective to acrolein.

$$C_3H_6 + O_2 \rightarrow C_3H_4O + H_2O \tag{3.3}$$

Iron oxide (Fe_2O_3), however, is very active, but non-selective, favouring the combustion route

$$C_3H_6 + \tfrac{9}{2}O_2 \rightarrow 3CO_2 + 3H_2O \tag{3.4}$$

However, if Fe is doped into the rutile lattice of Sb_2O_4 (which is an equal mixture of Sb^{3+} and Sb^{5+}) to replace the Sb^{3+}, then an active, highly selective catalyst is formed ($FeSbO_4$). In this case the main effect is on the stability of the oxygen in the lattice, which affects: (i) the surface reactivity and oxygen vacancy formation rate; and (ii) the rate of diffusion of oxygen vacancies into the bulk. The mixed phase material has a near optimum efficiency and is used commercially in a SiO_2 supported, multiply promoted form.

3.4 Support effects

The support can be used as a way of tailoring the performance of a catalyst, particularly if the active phase consists of small diameter particles. The area of the support affects the dispersion of the active phase, which then also tends to have a low surface area if the support is of low area. Impurities in the support can affect catalytic performance. Thus, if Rh is supported on highly pure SiO_2 it tends to produce alkanes/alkenes in synthesis gas (CO/H_2), whereas on standard silica, oxygenates can be formed. This is due to the presence of impurities of Fe and alkalis which can diffuse onto the metal phase and promote it for oxygenation

Some supports are reducible in reducing gases (TiO_2, for example) and the reduced form of the support is often labile and can diffuse onto the metal, giving marked effects on catalytic activity known as SMSIs (strong metal support interactions).

As described above, the nature of the interface and the dispersion of the metal alter the interaction with the support and to some degree these properties can be used to tailor performance.

II Practicalities: what is a catalyst?

A catalyst is a material that can increase the rate of a reaction, while not being consumed in the process. It is intimately involved in the reaction and is not a 'referee'. A referee does not get directly involved in the game, whereas a catalyst is very much involved, as a reagent, in the chemical reaction, but is regenerated at the end of the cycle. In this section we will consider the answers to the question 'what is a catalyst?' by examining what they are physically and chemically composed of, how they are made, what properties make good catalysts, the kinetics of catalytic processes and how the catalytic scientist can determine that the properties needed to make a good catalyst have been achieved.

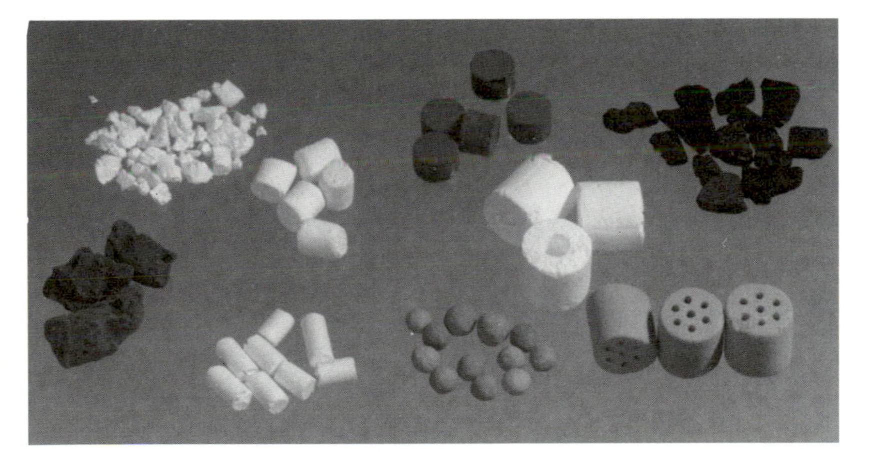

4 Catalytic materials and their preparation

4.1 Types of catalytic material

Catalysts can take a broad range of forms, and this is reflected in the range of materials produced, and used, by the chemical industry and by nature. Thus they can range from solid materials to biological enzymes, from gas phase molecules to liquid coatings on surfaces. In this section we will highlight the important types of applied industrial catalysts.

The most common materials used for catalysts are metals and oxides of various kinds. Metals are generally materials of very high surface energy and are therefore very active for many catalytic reactions. Indeed, for a wide range of reactions, particularly selective oxidations (for example, propene to acrolein), they are too active and their use results in poor selectivity for the desired product, mainly with combustion of the valuable reactant molecule. In these cases oxide catalysts are used (for instance, mixed oxides such as $FeSbO_4$ for propene oxidation), which, although usually less active than metals, give high selectivity for the partially oxidised product.

Most of these catalysts are not thermally stable in the high surface area form in which they need to be used, and so they are usually prepared as small particles adhering to a refractory support material – usually oxides such as alumina or silica. These help to reduce the rate of sintering, especially of metal particles (the effect of sintering was shown in Figs 1.2 and 2.11). The macroscopic form of the catalyst which is used is strongly dependent on the nature of the reaction, but a selection of types is shown in Fig. 4.1. The simplest form, which is often used in small-scale reactors (microreactors), is a powder. This can be used directly from the preparation, although this will often result in reactor blocking due to the presence of very fine powder. Thus it is usually made coarser by sieving, to use only large particles (say ~0.1 mm size), or more often the powder is compressed under pressure to form a large tablet, which is then broken up again, and the appropriate range of particle sizes is obtained by sieving. The latter process results in less waste catalyst, a narrower range of particle sizes and generally bigger particles.

In many cases shape is important for the best use of the material. Powder is generally useless in a large-scale, fixed-bed reactor, because the particles can: (i) get blown out of the bed; and (ii) block the bottom of the bed. The latter results in high resistance to gas flow and a large pressure drop across the catalyst. For a large-scale fixed-bed reactor, which may hold many tonnes of material in a 20 m tall tower, it is crucial that the catalyst is: (i) of a form that can allow significant gas flow with a low pressure drop; and (ii) resistant to crushing which would result in the reactor blocking with 'fines'. These properties are achieved by a variety of macroscopically hard shapes and some of these are shown in Fig. 4.1. They can be cylindrical pellets, pellets with holes in them, lumps, spheres or even be in the form of shaped metals with the catalyst coated on the surface. The latter would be useful when good thermal conductivity is essential to enable heat to get to, or from, the reaction mixture efficiently.

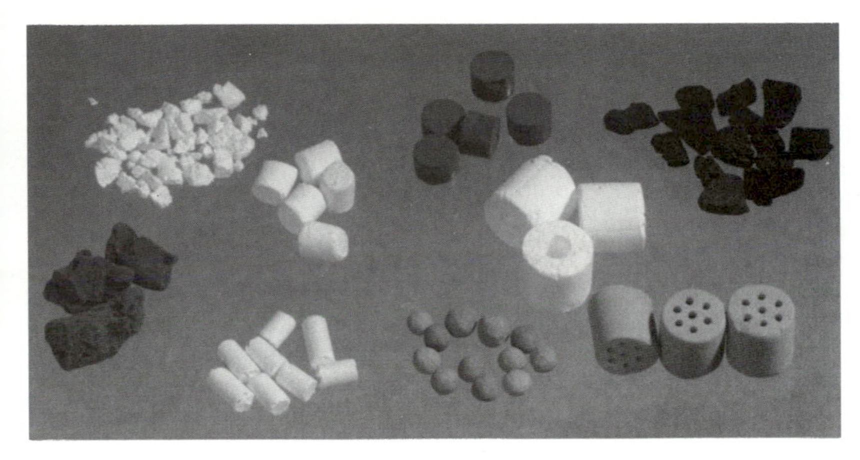

Fig. 4.1 Different forms of supported catalyst.

Some types of industrial reactor, however, require fines – very small particle sizes. Such reactors are fluidised bed reactors, which are described in more detail later on (Chapter 6). Here, the catalyst medium needs to be able to 'flow' in a gas stream, and small particle diameters (often ~50 μm diameter spheres) are essential for this. Under these circumstances powder may be used, but the choice of support is crucial (alumina, for instance, is very abrasive and may scour the reactor walls) and shape is important (microspheres may be essential).

A fundamentally important feature of a catalyst must be reiterated here, that is, its high surface area. A pellet of the type shown in Figs 1.1 and 4.1 earlier, can have a surface area of 100 m^2, that is, approximately the same area as all the pages of 15 copies of this book (~1500 pages). How is this possible in such a small amount of material with an apparent outside area of only 1 cm^2? The answer lies in porosity. The pellet mentioned contains a range of very small pores, as illustrated in Fig. 4.2. These can be divided into three arbitrary classes

The average size of pores in a material is approximately related to the surface area by the following relationship

$$SA \approx \frac{2V}{r} \qquad (4.1)$$

where *SA* is the surface area in the sample, *V* is the pore volume and *r* is the average pore radius. This relationship is derived simply from the geometry of an open cylinder, and the ratio of cylinder area to volume. Thus, if a sample has 50% porosity, a density of 3 g cm^{-3}, and a surface area of 100 m^2 g^{-1}, then the average pore size is ~3 nm. As shown in Fig. 4.2, this can be a gross oversimplification since a catalyst may contain a wide range of pore types.

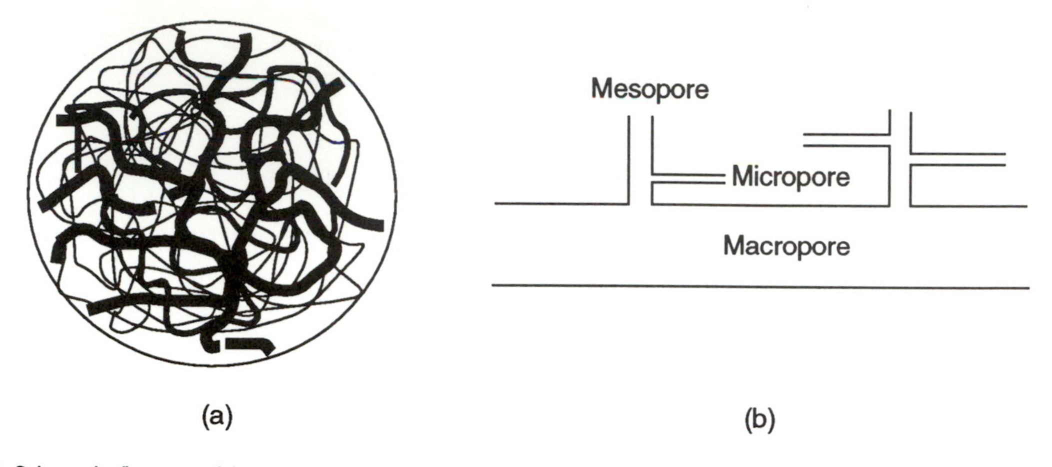

Fig. 4.2 Schematic diagrams of the pore structure of a catalyst showing (a) an interpenetrating array of different sized pores and (b) interconnection of macro-, meso- and micro-pores.

- Macropores – pores of > 100 nm diameter
- Mesopores – pores of > 2 nm and < 100 nm diameter
- Micropores – pores of < 2 nm diameter

Methods of assessing pore sizes and surface areas are described later (Chapter 6).

Macropores are essential for most catalysts to enable: (i) efficient penetration of liquids in some preparations, for instance, in incipient wetness methods (Section 4.4 below); and (ii) efficient flow of reactants to the smaller pores. The smaller types of pores are essential to the high surface area. Most catalysts are mesoporous with an interpenetrating array of variously sized pores. The highest surface area catalysts also have significant amounts of microporosity. For instance, as shown in Figs 4.3 and 4.4, zeolite catalysts are often very crystalline; they have very well-ordered, well-defined arrays of micropores, and can have surface areas of as much as 1000 m^2 g^{-1}.

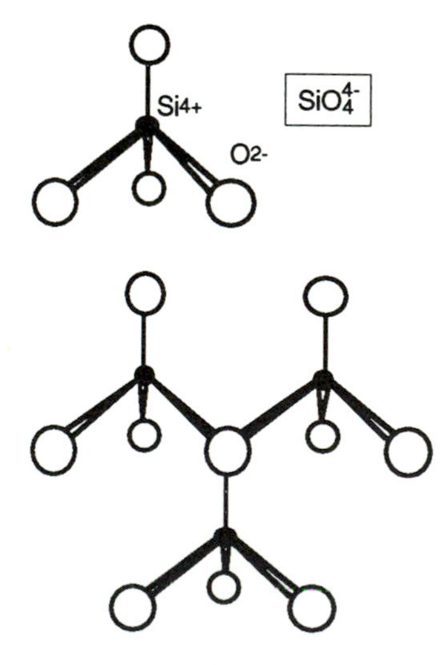

Fig. 4.3 Basic tetrahedral building block of silicon-based zeolites.

Zeolites have become very important materials for catalytic processing in recent times. Most zeolites are aluminosilicates based upon a simple tetrahedral building block (Fig. 4.3). Zeolites are mainly Si based, but the presence of small amounts of Al^{3+} in the framework has two main effects: (i) it distorts the lattice, resulting in new crystal structures and pore sizes/shapes; (ii) it is of lower charge, and therefore has to be balanced by the presence of another net charge cation which is exchangeable. The cation in the preparation is often an alkali metal such as Na which can be acid exchanged for a proton, creating a highly acidic site in the zeolite. From the basic tetrahedral unit the sodalite cage can be built (Fig. 4.4) and this, in turn, is the building block for a variety of other zeolite structures with different pore and internal cavity sizes. Some of these zeolites are naturally occurring (for example, faujasite), others are man-made (e.g. ZSM5). The latter is now widely used for the catalytic cracking of oil (see Chapter 7).

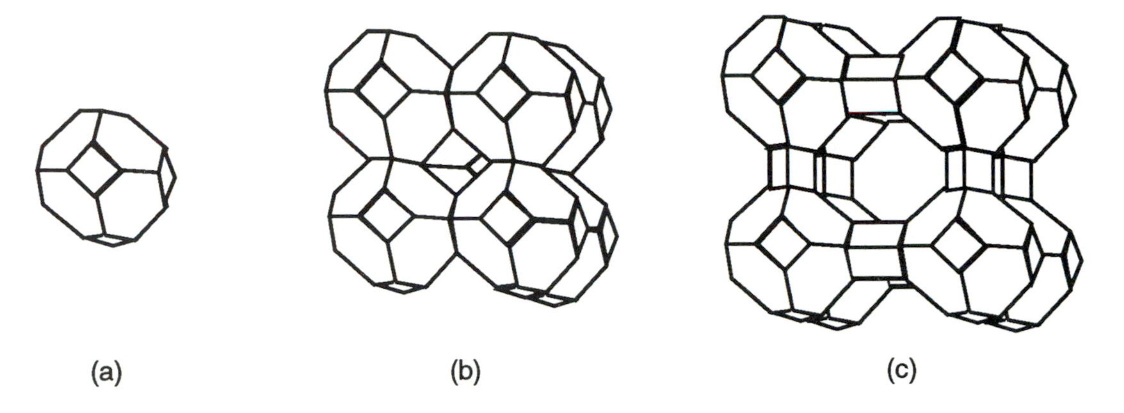

Fig. 4.4 Models of (a) sodalite cage, (b) sodalite and (c) Zeolite A. The intersects are cation positions.

They can also be used, in combination with metals incorporated into the cavities (often by metal ion exchange) for a wide range of catalytic conversions. Recently, a range of mesoporous zeolites have been invented, which may be of some use as catalyst supports with a very specific pore shape.

Having briefly listed the forms of catalyst that are used in industry and academia, we will now examine what parameters constitute an efficient catalyst for any particular reaction.

4.2 What makes a good catalyst?

There is a wide range of properties of a catalyst that are required in order for it to be an efficient one for any particular reaction. The foundations of catalyst production are the correct kinetic reaction parameters, that is, high product yield per unit time, and so, as shown in Fig. 4.5, the correct active phase is the basis of catalyst preparation. Without this piece in the catalyst development pyramid a successful material will not be obtained, whatever success there is with the other parameters. In order for the yield to be optimised, the catalyst should usually have a high surface area exposed to the reactant, although, for some reactions, we may want a modest surface area to avoid further reaction of an intermediate product. Items 3, 4 and 5 in Fig. 4.5 can be considered together. The longevity of a catalyst is a very important parameter in catalyst design, and one often ignored by academic workers; it is crucial for the cost effectiveness of a process.

Almost all catalytic processes decay in activity with time, as illustrated earlier in Fig. 3.1, and this is due to a variety of contributions as discussed in Chapter 3 in more detail. The cost effectiveness is determined by the integral under the time–yield plot. The bigger it is, the better. It is in this way that the vast majority of patents for new catalysts fail. It is relatively easy to invent a very active catalyst, but such materials usually decay in activity very quickly. The decay is caused by a variety of factors, of which three dominate.

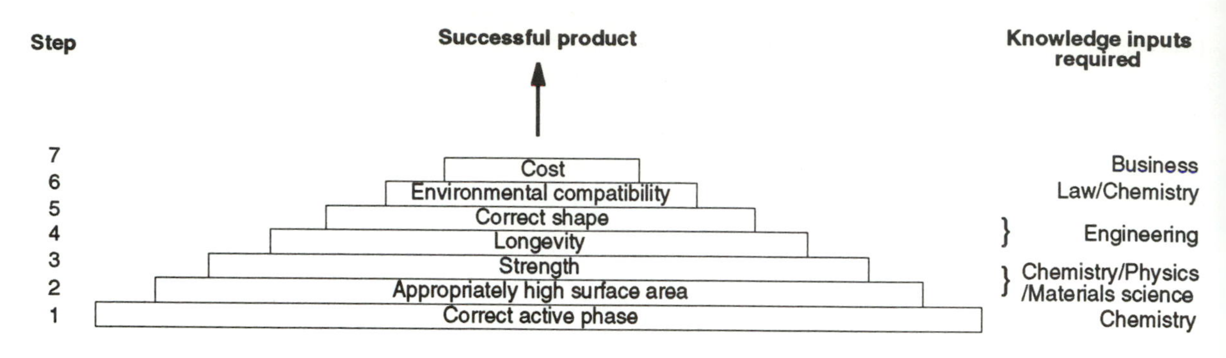

Fig. 4.5 Pyramid of expertise involved in catalyst development.

(a) sintering – loss of surface area
(b) attrition – powdering of catalyst, blocking reactor, reducing gas flow
(c) poisoning – loss of activity per unit area

Point (a) is minimized by the correct preparation procedure in steps 1 and 2 shown in Fig. 4.5. Point (b) is ameliorated by paying attention mainly to step 3, but step 5 also plays a role. Poisoning [point (c)] can be avoided by using a guard bed at the front of the reactor to trap the poison, and by making use of a great deal of gas 'scrubbing' and clean-up upstream of the reaction bed. It is essential to pay attention to point 5 in order to obtain good gas and heat flow characteristics down the reactor, and to minimise crushing of material at the bottom (which may have a significant weight of catalyst bearing down on it).

Step 6, shown in Fig. 4.5, has become of great significance in recent years, and rightly so. The catalyst must be environmentally compatible in a variety of ways:

(a) it must minimise toxic by-products in the reaction;
(b) it must not release toxic materials itself to the environment (e.g. heavy metals) during use;
(c) it must be in a low toxicity form if being dumped after use, or better still, it should be either regenerable or recyclable.

All of these factors are essential for a successful catalyst, both in terms of current and future legislation limits and, perhaps more importantly, in terms of the fact that chemistry in the future must be seen as a clean, modern technology, not as a science for the poisoning of mankind. Ultimately these factors are also much more cost effective on the global scale.

On the last step in the pyramid, a process catalyst will fail if the costs of producing it are high compared with the selling price of the product, which in turn relates to competition in the market place, both for the catalyst, for the process, and for the product of the catalysis itself. A very high price can be obtained for a catalyst which can facilitate a new reaction or process. In general terms, a catalyst can be classed as an 'effect chemical' or a 'higher added value' product. This puts catalysts in the same category of materials as pharmaceuticals, for which the profit margin is often more than 30%.

In light of Fig. 4.5 it can be appreciated that catalysis is an interdisciplinary subject, requiring knowledge over a range of fields, from fundamental and applied chemistry, through materials science, engineering and business/legal inputs. Successful catalysts can only be produced by the close interaction of all these disciplines and the best organizations in the world at catalytic innovation have research groups set up that involve all of these. The weaker companies who are doing badly at catalytic innovation have resisted such ideas for some considerable time. In the following section approaches to 'catalyst design' will be considered before covering methods of catalyst synthesis in section 4.4.

4.3 Catalyst design methods

Figure 4.6 summarises approaches to the integrated design of catalysts. The methodology for most of the era of catalyst production up till the present has been somewhat 'empirical'. The latter word is one which is often used disparagingly, but this is misplaced, since the word is literally meant to describe an experimental approach to problem-solving, an approach upon which most of science is built. However, such an approach can also be rather 'hit and miss', relying heavily on chance findings and serendipity. There was very little input of detailed theoretical understanding of catalysis, although a deep understanding of synthetic chemistry and chemical properties was essential for the evolution of catalysts in this way. Rather than 'catalysts by design' this approach is better classified as catalyst development.

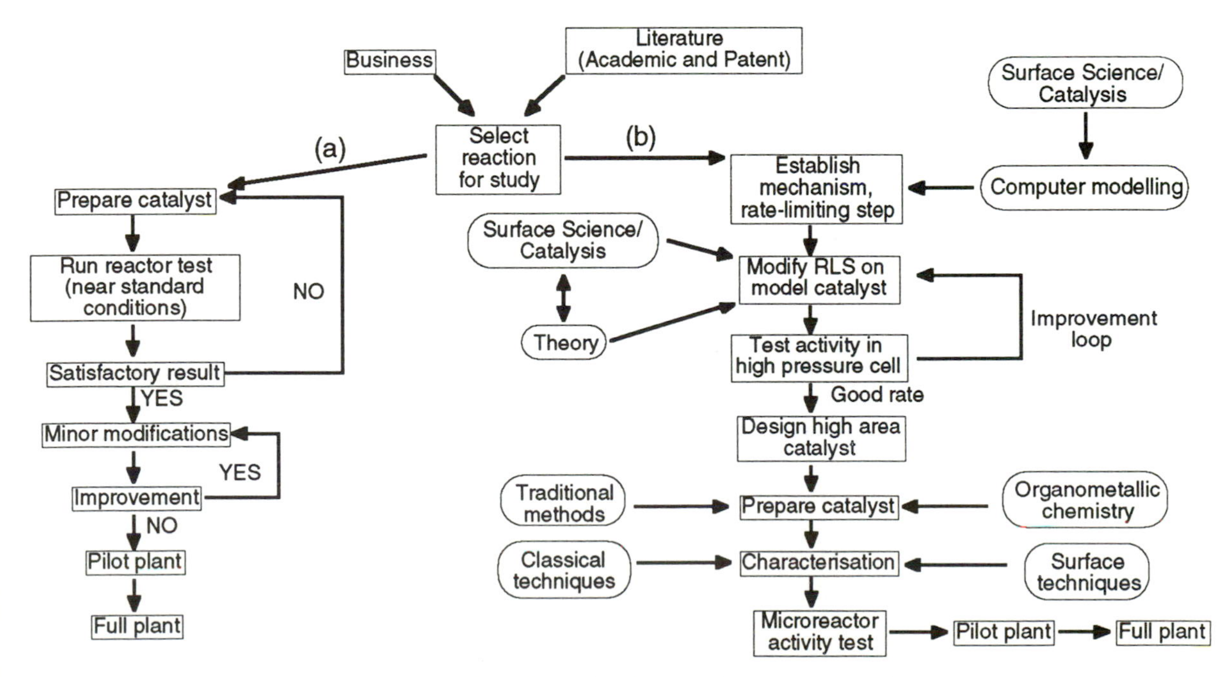

Fig. 4.6 Catalyst development flow chart; (a) traditional method, (b) modern integrated method.

With advancing theoretical understanding of catalysis, and the development of computational power and so-called 'expert systems', a new era of 'catalysis by design' is dawning and this approach is also outlined in Fig. 4.6. Here the important point is that targets can be better identified. Let us give an example of this approach, and let us use the case of ammonia synthesis catalysis. The catalyst for this process is an excellent one, evolved over 90 years of experiments, but not essentially very different from that discovered near the beginning of this period. The approach then was entirely empirical, many reactors were used 24 hours a day at BASF to 'screen' materials (ores, for instance) from around the world for the most efficient catalyst. Eventually, Swedish magnetite was identified as the best, it was analysed, and synthetic analogues were produced which contained the essential promoting additives (most importantly potash and alumina). Let us pose a question; how do we make an even better ammonia synthesis catalyst? The current catalyst has been thought to be so good, and to have been around so long, that it was impossible to invent a better one. This is, of course, scientific nonsense, and probably impedes progress in this area. To make a better catalyst we have to look to the fundamental limitations in the reaction and this requires literature input from the surface science and catalysis areas. It is thought that the dissociation of N_2 molecules is slow, and that, in turn, this is limited by the availability of surface sites due to blockage by N adatoms.

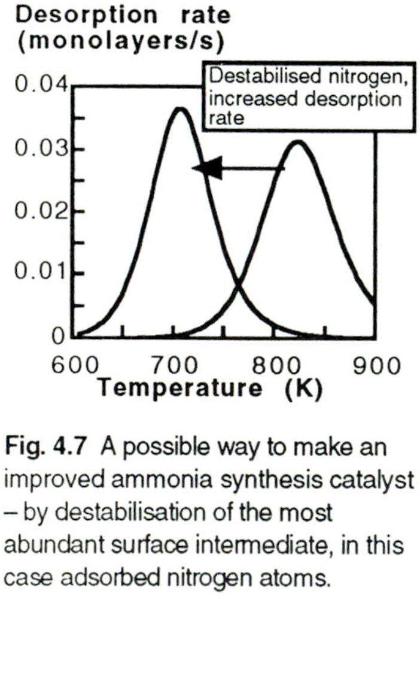

Thus nitrogen atom recombination and desorption might actually aid synthesis due to reduced self-poisoning [a lower steady state N(a) coverage]. Thus in designing a new catalyst we would look to a material which has a lower desorption temperature for nitrogen atom recombination than at present (see Fig. 4.7). Verification that this would indeed be effective lies in the so-called 'microkinetic modelling' method, in which the experimentally determined kinetic and thermodynamic parameters for each step in the reaction are put into a computer, with the mass balance and flow characteristics for a particular reactor. The effect of changing any individual parameter on output can then be assessed. Only changes in those factors associated with the rate determining step should have a significant effect on the production rate, although if such a step is made more efficient another step may become rate determining. Such modelling then feeds back into the catalyst design. Modern synthetic and experimental skills are crucial, then, to produce the desired effect in a real material, but these steps are the essence of catalyst design in the 21st century. Returning to the example cited, the target might be for a material with reduced N(a) stability and this can then be explored for other metals, or for other additives to the current catalyst. However, care must be taken here because very often any individual component has a multiplicity of roles and effects. For example, in current catalysts potash acts as an activity promoter, but also interacts strongly with alumina and may block sites there that are active for NH_3 binding and decomposition. A new generation of catalysts has recently appeared which uses Ru as the main metal component and which indeed appears to have reduced N(a) stability. A further major advantage of the microkinetic modelling method is that it may indicate new areas of temperature, pressure and reactant ratio which may be beneficial for

Fig. 4.7 A possible way to make an improved ammonia synthesis catalyst – by destabilisation of the most abundant surface intermediate, in this case adsorbed nitrogen atoms.

the reaction. A great limitation of current exploration methods is often that they are carried out under the fixed conditions that are best for current catalysts, but may not be appropriate to new generations of materials. In this respect catalytic innovation is often limited by engineering considerations and the cost of building new plants. Thus, it must be remembered that the catalyst design function is step 1 in Fig. 4.5 and must be integrated with the other requirements of any particular company.

4.4 The making of catalysts

Catalysts can be made by many of the variety of ways available to chemistry, but here we will restrict ourselves to a summary of the most commonly used methods.

Impregnation

A supported material may be prepared by starting with a high area, porous support, such as alumina, and dosing it with the active phase precursor. This impregnation can be achieved in a variety of ways. It can be done by adsorption of an active phase from solution, that is, by adding the support to a solution of the active ingredient. This method can be modified by precipitation impregnation, in which the active phase is precipitated onto the support by, for instance, a sudden pH change in the mixture, or by chemical reaction. The catalyst then has to be separated by filtering, and much of the solution may be wasted.

In many ways a more controllable method is the 'incipient wetness' technique. Here, the active phase is added in solution form to the dry powdered support until the mixture goes slightly tacky. This occurs when the pores of the support are filled with the liquid.

In all these cases the impregnated material is finally dried and then usually calcined (heated in an oven) to some specific temperature which decomposes the precursor active phase to produce either the final active phase, or something close to it.

Slurry precipitation

In this case the catalyst is directly produced from the active components by mixing. If we take an example of $FeSbO_4$ catalysts for propene oxidation or ammoxidation, these can be prepared by warming $FeNO_3.9H_2O$ to 60 °C, at which point the water of hydration of the salt is released and the nitrate dissolves in it to form a solution. Sb_2O_3 is then added, in powder form, to this solution, and the pH is raised by the addition of aqueous ammonia. The final catalyst is produced by drying and calcining to very high temperature which enables ionic interdiffusion to produce the rutile-type $FeSbO_4$.

Co-precipitation

Here the precursor components are mixed and then co-precipitation occurs at a particular pH by addition of other salts, acids or bases. A good example of this method is the production of methanol synthesis catalysts. Here, an intimate mix of nitrates of copper, zinc and aluminium are co-precipitated by rapid addition of Na_2CO_3. The gel which is produced is aged (left to change chemical and physical structure slowly), has to be washed to remove Na ions which would otherwise change the selectivity to methanol, and finally has to be dried and calcined. The final active product is a mix of copper oxide, zinc oxide and alumina as the support phase.

Fusion

The most notable example of this kind of preparation is the ammonia synthesis catalyst for which mainly magnetite, together with small amounts of potash, lime and alumina, are fused at high temperature (~1800 K) in a melt. This is then cooled and broken into somewhat irregular lumps of approx. 5 mm diameter. This produces a very low area material with no pore structure. This is only developed when the iron oxide is reduced *in situ* in the reactor to produce the metallic form, which then has pores where the hydrogen penetrated the material to reduce it.

The Raney method

Here, two metals are mixed, one being leachable in an acid or alkali medium. Thus, Raney nickel is produced from a NiAl alloy, from which the Al is leached with alkali. This leaves a highly porous, highly reactive, high surface area form of the metal, but if done carefully it also leaves some alumina which helps maintain the integrity of the catalyst during use. Such catalysts are used in hydrogenation of unsaturated vegetable oils to produce more 'spreadable' margarine, and in this way much of the food we eat has actually passed through an industrial catalyst first (further details in Chapter 9).

Physical mixing

Sometimes good catalysts can be produced in this way. Dry powder of the individual components (e.g. Fe_2O_3 and Sb_2O_3 in the case of the catalyst for propene oxidation, or magnetite, potash and alumina for ammonia synthesis catalysts) are mixed and calcined to high temperature. This can often form a mixed phase by ionic interdiffusion, or can form 'necking' between individual particles. There is some evidence that a synergistic effect is obtained from the mere act of mixing. This is often put down to the ability of one phase to adsorb one reactive component, whereas the other cannot, but this reactant diffuses onto the latter where it reacts to make the product. In general such mixes, although often having surprisingly good characteristics, do not match up to those produced by more sophisticated methods which give a better intimacy of the component ions or elements.

Washcoating

Although mentioned last here, this method is the largest scale industrial catalyst preparation nowadays because it is used in car catalyst production. The monolith support (see Fig. 8.3) is the component of these catalysts which gives good thermal shock (rapid heating) resistance, while allowing the exhaust gases to flow at very high velocity. The active phase only coats the outside geometric area of this honeycomb, but a high area is produced by first coating the honeycomb with a so-called 'washcoat' layer, which is basically a highly porous alumina, with a variety of other minor additives for resistance to sintering. The final catalyst is then produced by dipping the monolith into a mix of the active components (esp. Pt, Pd and Rh salts), drying and calcining.

Pelleting

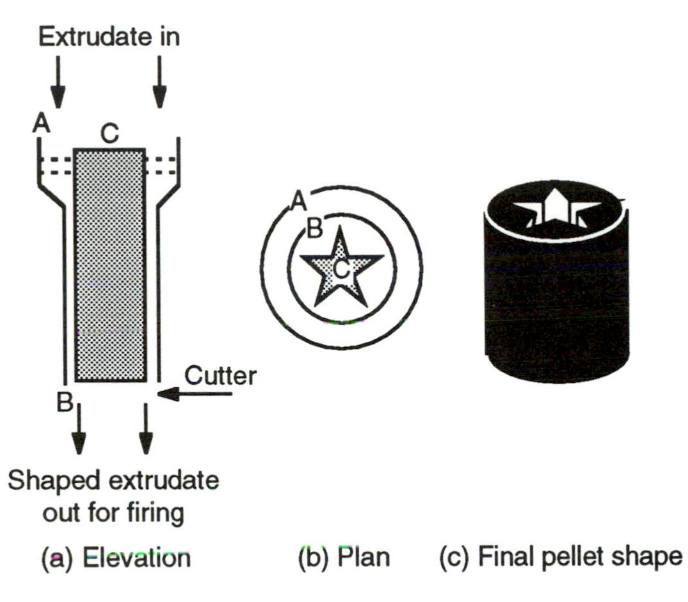

Fig. 4.8 An example of a catalyst pellet produced by extrusion.

The final step in the production of many industrial, heterogeneous catalysts, is pelleting to produce the macroscopic form which is used in the plant, examples being shown in Fig. 4.1. This is usually done in two ways. One is high pressure compaction in a pelleting machine, the other is by extrusion of a plasticised version of the catalyst powder. This is in the form of a paste, which is forced through a former (Fig. 4.8), in the simplest case producing a cylindrical rod of extrudate which can then be dried/fired to harden the material. A variety of shapes can be produced, however, by appropriate design of the extrusion formers.

5 Catalytic activity and selectivity

5.1 Kinetics of catalytic conversion

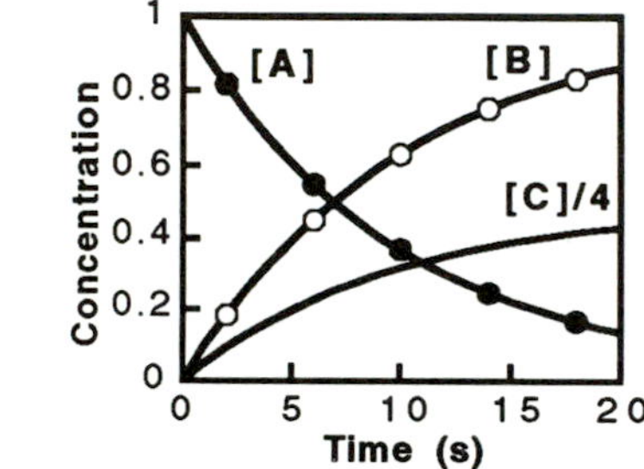

Fig. 5.1 The exponential dependence of reactant and product concentration upon time for a simple, first-order process.

$\frac{-d[A]}{dt}$ = activity
(Units: moles/dm^3/s)

$\frac{d[B]}{dt}$ = activity to product B

$\frac{d[C]}{dt}$ = activity to product C

Fractional conversion of A

$$C_A = \frac{[A]_0 - [A]}{[A]_0} \qquad (5.2)$$

where A_0 is the starting concentration of A

Selectivity to B for parallel reaction

$$A \rightarrow B, \quad A \rightarrow C$$

$$= \frac{B}{[B]+[C]} = \frac{[B]}{[A]_0-[A]} = \frac{[B]}{[A]_0.C_A} \qquad (5.3)$$

Selectivity to C

$$= \frac{[C]}{[B]+[C]} = \frac{[C]}{[A]_0-[A]} = \frac{[C]}{[A]_0.C_A} \qquad (5.4)$$

The most important feature of a catalyst is how well it does the job of converting reactants into products, and this is measured by two parameters – activity and selectivity. The activity of a catalyst is defined as the rate of consumption of reactant, although the activity to a particular product can also be specified, whereas selectivity is the fraction of the total products which a particular substance represents. Figure 5.1 illustrates these terms for a so-called 'batch' reaction in a closed pot, where reactant A is broken down into products B and C ($A \rightarrow B + 2C$).

A further term, known as specific activity, can be written as follows

$$A_{S_V} = \frac{-1}{S_V} \cdot \frac{d[A]}{dt} \qquad (5.1)$$

where S_v is the total number of active sites per unit volume. This number then indicates which material may be best in terms of activity per unit volume. This is often the most important representation industrially, where reactors have a fixed volume to be filled with catalyst. However, specific activity can also be expressed per unit weight of catalyst.

Instead of expressing this quantity in per unit volume terms it can be expressed per unit surface area.

$$A_{S_A} = \frac{-1}{S_A} \cdot \frac{d[A]}{dt} = \frac{1}{b.S_A} \cdot \frac{d[B]}{dt} \qquad (5.5)$$

where b is the stoichiometric number of the conversion of one mole of A to B (one in the example given here). S_A is the number of active sites per unit area of catalyst surface. An alternative expression gives the rate in terms of molecules reacted site^{-1} s^{-1}, called turnover frequency, T, which can be expressed as follows,

$$T = A_{S_V}.N_A \qquad (5.6)$$

where N_A is Avogadros number. T is a widely used number in the academic catalytic literature, but is a rather poor parameter to represent effectiveness in catalysis. First, it is quite possible to have a catalyst with a very high turnover number which is of little use in a catalytic reactor because it has a very low surface area per unit volume, that is, very few sites per unit volume. Thus T can be high and total activity can be low.

These are the kinetic parameters most commonly used to describe a catalytic reaction and an example of the kind of graph often presented to describe results in this field is given in Fig. 5.2. However, if a basic mechanistic understanding of catalysis is the aim of much of catalyst research, then more detailed consideration has to be given to the many catalytic parameters which are involved in the processes shown in Fig. 1.5 and this begins with adsorption, the first surface step of catalytic turnover.

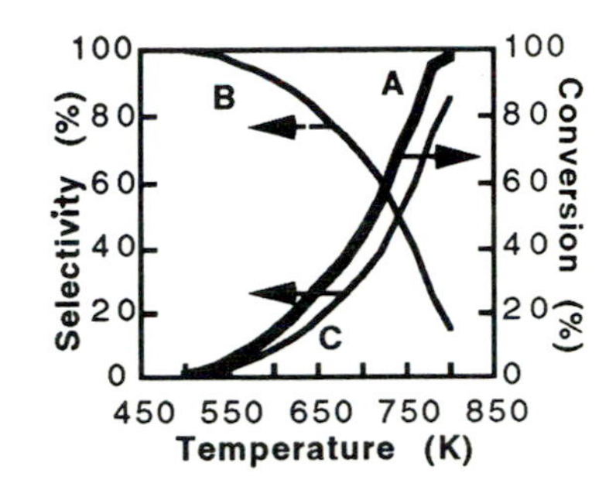

Fig. 5.2 A typical example of a plot of catalytic kinetic properties determined in a flow reactor for a reaction with two possible products in a sequential process of type A → B → C.

Adsorption

One of the early pioneers of this area, and perhaps the most significant figure in surface chemistry, was Irving Langmuir who derived the following relationships. If we imagine a molecule hitting a surface, then it is likely that there are individual sites at which it can 'stick', that is, interact with the surface by bonding with it.

$$A_g + * \rightarrow A_a$$

This is called adsorption and the rate is given by

$$\frac{-d[A_g]}{dt} = S_0 \cdot P \cdot Z. (1-\theta) = k_a(1-\theta) \cdot P \qquad (5.7)$$

Here S is the sticking probability of the gas on the clean surface, P is the pressure of gas, Z is the Knudsen collision factor and θ is the fractional blocking of the surface by adsorbate. The asterisk represents the site on the surface for adsorption, and subscripts 'g' and 'a' refer to gas phase and adsorbed states, respectively. When the surface is filled with adsorbate (θ=1) the surface is unreactive and the adsorption rate is zero.

Adsorption – desorption equilibrium

Equation 5.6 is a rate equation, but it is possible to derive the surface coverage by adsorbate when there is a dynamic equilibrium present, that is, when the adsorbate A is capable of desorbing again and returning to the gas phase. Then we need to consider also the desorption rate,

$$A_a \rightarrow A_g + *$$

$$\frac{-d[A_a]}{dt} = k_d \cdot \theta \qquad (5.8)$$

this is directly proportional to the surface concentration θ, and the proportionality constant is the desorption rate constant. Thus, at equilibrium

$$\frac{-d[A_g]}{dt} = \frac{-d[A_a]}{dt} \qquad (5.9)$$

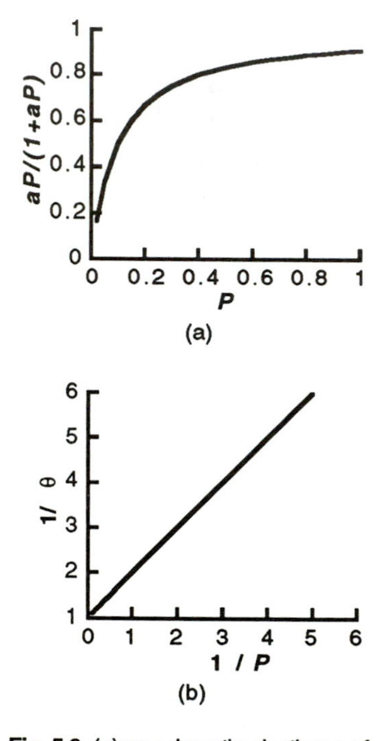

Fig. 5.3 (a) an adsorption isotherm of the Langmuir type with $a = 10$ and (b) the linearised form of this plot.

therefore, it follows that,

$$\theta = \frac{aP}{1+aP} \quad \text{or} \quad P = \frac{\theta}{a(1-\theta)} \quad \text{or} \quad \frac{1}{\theta} = \frac{1}{aP} + 1 \tag{5.10}$$

where a is the adsorption equilibrium constant.

Data can be obtained for the equilibrium coverage of the surface, at a particular temperature, over a range of pressures and then a linearised isotherm can be plotted as shown in Fig. 5.3, using the reciprocal form of the Langmuir isotherm shown on the right of Eqn 5.10. The slope of this is $1/a$, the reciprocal of the equilibrium constant, from which the heat of adsorption can be determined

$$a = \frac{k_a}{k_d} = \frac{A_a}{A_d} \cdot \frac{e^{(-E_a/RT)}}{e^{(-E_d/RT)}} = \frac{A_a}{A_d} \cdot e^{(-\Delta H_a/RT)} \tag{5.11}$$

The A factors are kinetic pre-exponentials relating to surface collision and bond breaking and ΔH_a is the heat of adsorption (equal to the difference in adsorption and desorption activation energies).

Simple reaction

Real surface reactions contain the above elements, but also a reaction event which yields products. Thus, at the simplest levels there are now four steps involved

$$A_g + * \rightarrow A_a \quad \text{(step 1)}$$
$$A_a \rightarrow A_g + * \quad \text{(step 2)}$$
$$A_a \rightarrow B_a \quad \text{(step 3)}$$
$$B_a \rightarrow B_g + * \quad \text{(step 4)}$$

if we consider the process to be rate limited by step 3 (4 is fast) then

$$\frac{-\,d[A_g]}{dt} = \frac{d[B_g]}{dt} = k_3 \,.\, \theta_A \tag{5.12}$$

where θ_A is the coverage of adsorbate A on the surface (since B desorbs fast its coverage is assumed negligible).
In turn,

$$\frac{d\theta_A}{dt} = k_1 P_A(1-\theta_A) - k_2\theta_A - k_3\theta_A = 0 \tag{5.13}$$

The latter is the steady state assumption that θ_A is invariant with time and so

$$\frac{d[B_g]}{dt} = k_1 P_A - \theta(k_1 P_A + k_2) \tag{5.14}$$

Further, from Eqn (5.13)

$$\theta_A = \frac{k_1 P_A}{k_1 P_A + k_2 + k_3} \tag{5.15}$$

and substitution into Eqn 5.14 yields the following

$$\frac{d[B_g]}{dt} = \frac{k_1 k_3 P_A}{k_1 P_A + k_2 + k_3} = \frac{k_3 K P_A}{1+KP} \tag{5.16}$$

where K is a complex rate constant ($= \frac{k_1}{k_2+k_3}$). This is known as the Langmuir Equation in catalysis (not to be confused with the Langmuir isotherm). The same derivation is obtained if one assumes pre-equilibrium in steps 1 and 2 (that is, both go much faster than step 3) and if use is made of the Langmuir isotherm for determining θ. However, in that case, since step 3 is so very slow compared with 1 and 2, (k_3 is very small), then K in Eqn 5.16 above is the adsorption equilibrium constant ($= \frac{k_1}{k_2}$). Equation 5.16 is a generally applicable one in science, applying to many three-step reactions and to pre-equilibrium situations, and is widely known in the biochemistry of enzymatic catalysis as the Michaelis–Menten equation. K above is then simply the reciprocal of the so-called Michaelis constant.

As Fig. 5.4 shows, a characteristic of this equation is that the reactant pressure dependence of the rate changes from first order at low pressures to zero order at high pressures as the surface becomes saturated with adsorbate and $\theta \sim 1$.

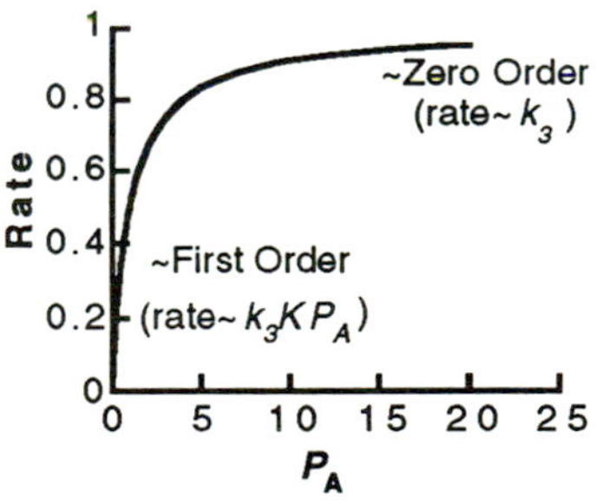

Fig. 5.4 The pressure dependence of reaction rate for a system obeying Langmuirian kinetics.

Bimolecular reactions

For a reaction A + B $\rightarrow$ C, the full mechanism should be written in single steps as follows:

$$A_g + * \rightleftharpoons A_a \quad \text{(step 1)}$$
$$B_g + * \rightleftharpoons B_a \quad \text{(step 2)}$$
$$A_a + B_a \rightarrow C_a \quad \text{(step 3)}$$
$$C_a \rightarrow C_g \quad \text{(step 4)}$$

Here, for brevity, both adsorption steps have been shown as being at equilibrium. We must note, first, that the Langmuir isotherm is modified for such competitive molecular adsorption as follows, the competitive term appearing in the denominator

$$\theta_A = \frac{aP_A}{1+aP_A+bP_B} \quad \text{and} \quad \theta_B = \frac{bP_B}{1+aP_A+bP_B} \tag{5.17}$$

where a and b are the adsorption equilibrium constants for species A and B.

The rate is limited by step 3 and so the rate equation has the following form and is known as the Langmuir–Hinshelwood equation.

$$\frac{d[C_g]}{dt} = k_3\theta_A\theta_B = \frac{k_3.a.b.P_A.P_B}{(1+aP_A+bP_B)^2} \tag{5.18}$$

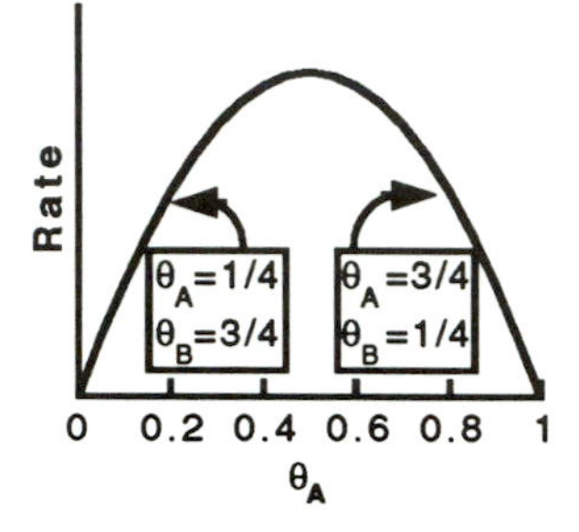

Fig. 5.5 Rate dependence on adsorbate coverage for a bimolecular Langmuir–Hinshelwood reaction.

Figure 5.5 shows some consequences of this relationship. At low pressures the equation is essentially first order in each reactant. When one reactant is weakly adsorbed, the terms involving that co-adsorbate are eliminated from the denominator (small adsorption equilibrium constant) and the reaction effectively becomes poisoned by a high coverage of the stronger binding adsorbate. If B is weakly adsorbed and A very strongly [that is, the denominator in Eqn 5.18 above reduces to $(aP_A)^2$] then Eqn 5.18 can be written approximately as

$$\frac{d[C_g]}{dt} = \frac{k_3.bP_B}{aP_A} \tag{5.19}$$

that is, there is negative first order in A and first order dependence on the more weakly held species. If both adsorbates are weakly adsorbed then the denominator in Eqn 5.18 reduces to unity and the reaction becomes first order in both components (coverage of both low).

$$\frac{d[C_g]}{dt} = k_3.a.b.P_A.P_B \tag{5.20}$$

A variant of this treatment yields the so called Eley–Rideal equation for the situation where one species is not adsorbed at all. Here, A is adsorbed on the surface, whereas B reacts by impulsive collision with adsorbed A at the surface, hence is simple first order in B, but Langmuir form in A.

$$\frac{d[C_g]}{dt} = \frac{k_3.a.b.P_A.P_B}{1+aP_A} \tag{5.21}$$

Reactions of this type have been much searched after in the last few decades, with little good evidence that they exist, except in special cases. Of course they must happen, but often the cross-section for reaction and the reaction probability are much lower than for bimolecular surface reactions, due to the catalytic effect on bond reorganization, and intermediate stabilisation in the latter case.

Temperature dependence of the rate

Very often catalytic reactions show a turnover in reaction rate at a certain temperature and this is due to the nature of the surface involved in the catalysis, which can itself be very different at different temperatures. If we use as an example a very simple catalytic reaction, namely, the oxidation of carbon monoxide, then the rate of reaction shows a maximum with temperature (Fig. 5.6). This is because the surface is dominated (and self-poisoned) at low temperature by CO molecules and at a high temperature by oxygen atoms (but not poisoned). Since the surface reaction is of the Langmuir–Hinshelwood type then the rate can be written as follows (Eqn 5.22), and since at the temperature extremes, the coverage of one component is low, so there are low rates there.

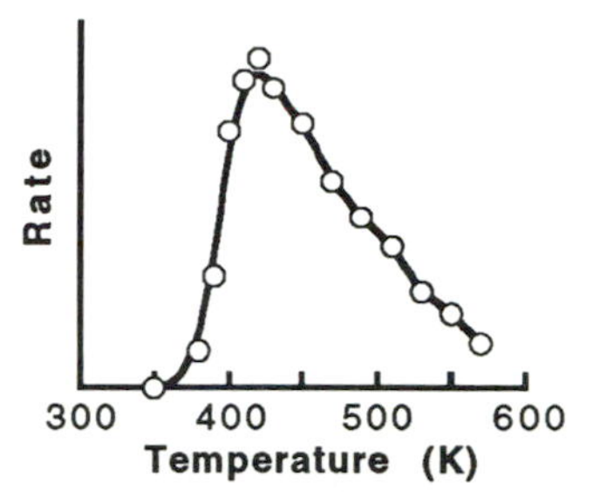

Fig. 5.6 Plot of steady state rate of CO_2 production from a molecular beam of CO and O_2 impinging on a Rh(110) crystal at several crystal temperatures. [From M. Bowker, Q. Guo and R.W. Joyner, Surf. Sci. 280 (1993) 50.]

$$R = k[\mathrm{CO(a)}][\mathrm{O(a)}] \tag{5.22}$$

In the middle temperature range there is a reasonable coverage of both components. In general terms, for a simple bimolecular surface reaction between species that compete for sites, but for which one is a much stronger adsorber, the rate is proportional to the coverage and can be written as follows.

$$\text{Rate} = k\theta(1-\theta) \tag{5.23}$$

This can be expanded by use of the Langmuir isotherm to represent coverage θ.

$$\text{Rate} = k.\frac{aP}{1+aP}\left(1 - \frac{aP}{1+aP}\right) \tag{5.24}$$

Note that the denominator here is simpler than that shown in Eqn 5.17, since one adsorbate is considered to be a weak adsorber; Eqn 5.24 simplifies to:

$$\text{Rate} = \frac{kaP}{(1+aP)^2} \tag{5.25}$$

At low pressure the denominator is unity and so this can be rewritten with rate constants expanded in terms of the individual Arrhenius expressions,

$$\text{Rate} = kaP = \frac{A\,\mathrm{e}^{(-E/RT)}.A_\mathrm{a}\,\mathrm{e}^{(-E_\mathrm{a}/RT)}.P}{A_\mathrm{d}\,\mathrm{e}^{(-E_\mathrm{d}/RT)}} \tag{5.26}$$

$$= \frac{A.A_\mathrm{a}}{A_\mathrm{d}}\,\mathrm{e}^{-\{(E+E_\mathrm{a}-E_\mathrm{d})/RT\}}.P \tag{5.27}$$

where the subscripts a and d refer to adsorption and desorption steps, respectively. Since $E_\mathrm{a}-E_\mathrm{d} = \Delta H_\mathrm{a}$, then

$$\text{Rate} = A_{app}\, e^{-(E_{app}/RT)}.P \qquad (5.28)$$

Since E_{app}, the apparent activation energy which would be experimentally determined, is equal to $E+\Delta H_a$, then if the heat of adsorption (a negative quantity) is bigger than the true reaction activation energy, then the rate will decrease with increasing temperature, since the exponential is a positive term which then decreases with increasing temperature. This explains the high temperature part of the curve of Fig. 5.6. However, it does not explain the lower temperature part. Then the assumption above, that the denominator in Eqn 5.25 is unity breaks down because the adsorbate coverage is high. Then Eqn 5.25 reduces to the following, since aP is much greater than unity

$$\text{Rate} = \frac{k}{aP} \qquad (5.29)$$

and so, expanding the terms

$$\text{Rate} = A\, e^{(-E/RT)}\, \frac{A_d\, e^{(-E_d/RT)}}{A_a\, e^{(-E_a/RT)}} \qquad (5.30)$$

$$= \frac{A.A_a}{A_d P}\, e^{-\{(E+E_d-E_a)/RT\}} = \frac{A_{app}}{P}\, e^{-(E_{app}/RT)} \qquad (5.31)$$

Now, in the terms discussed above, E_{app} will be a positive quantity, since it is now equal to $E-\Delta H_a$, and so the rate will increase with temperature. Note that Eqn 5.28 has a positive pressure dependence and Eqn 5.31 has a negative pressure dependence, which indeed mimics the reaction order in CO at the two temperature extremes.

5.2 Competing products: the selectivity problem

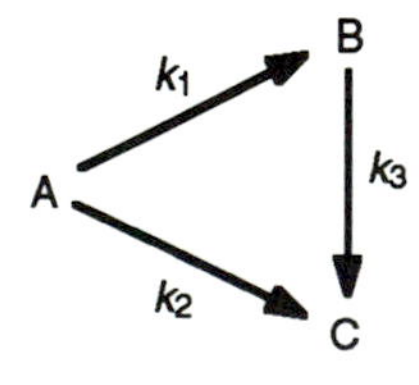

Scheme 5.1 Competing products from reactant A.

In general, in catalysis, more than one product is formed from any particular reaction, and only one of these is the desired product. The objective of the catalytic scientist, then, is to optimise the reaction for that desired product and this is achieved by the appropriate catalytic design. We can consider a simplified reaction scheme, with only three steps and two products as shown in Scheme 5.1. This is typical of many oxidation processes, where the undesired product (C) is the thermodynamically favoured one. For instance, in the selective oxidation of propylene to acrolein the thermodynamics are as follows, showing the enormous favourability of combustion over selective oxidation.

$$C_3H_6 + O_2 \rightarrow C_3H_4O + H_2O \qquad \Delta H^o = -336 \text{ kJ mol}^{-1} \qquad (5.32)$$

$$C_3H_6 + \tfrac{9}{2}O_2 \rightarrow 3CO_2 + 3H_2O \qquad \Delta H^o = -2060 \text{ kJ mol}^{-1} \qquad (5.33)$$

The catalyst has to play the role of intercepting the reaction at the first stage, which is not easy, and it is impossible to achieve 100% selectivity to acrolein. Nevertheless, modern catalysts have been designed to have very high selectivity for this reaction (>95%) at high conversion (~95%).

Selectivity has been defined earlier (Section 5.1) and here we will look only at the effects of differing values for the rate constant on the reaction. Clearly, if k_2 and k_3 are small, then it is relatively easy to make product B in high selectivity. If k_3 is small, but the others are significant then as A is converted the results shown in Fig. 5.7 are obtained, that is, selectivity is constant with conversion, at the value dictated by the ratio of rate constants. However, if k_3 is significant, then as the conversion proceeds, ultimately all of the reactant is converted to the undesired product. This is the situation for most real oxidation reactions, an example being given here for ethylene oxidation on an Ag catalyst (Fig. 5.8). It is for this reason that many industrial heterogeneous selective oxidations run at low conversion – to avoid over-oxidation of the product. In the case of ethylene epoxidation the reaction is run at ~30% conversion, with recycling of the unused hydrocarbon to the front of the reactor.

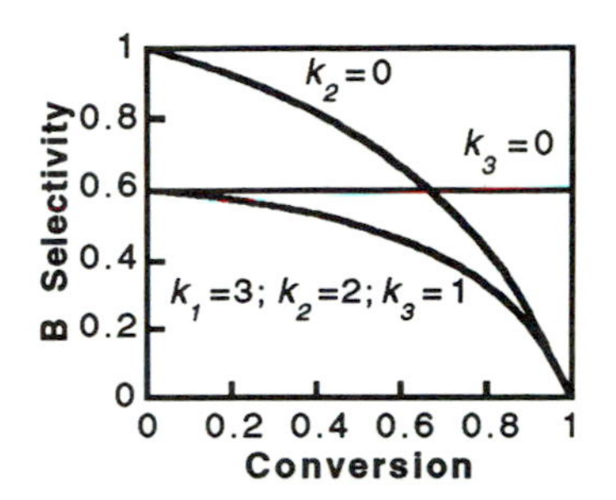

Fig. 5.7 Generalised forms of selectivity dependence upon conversion.

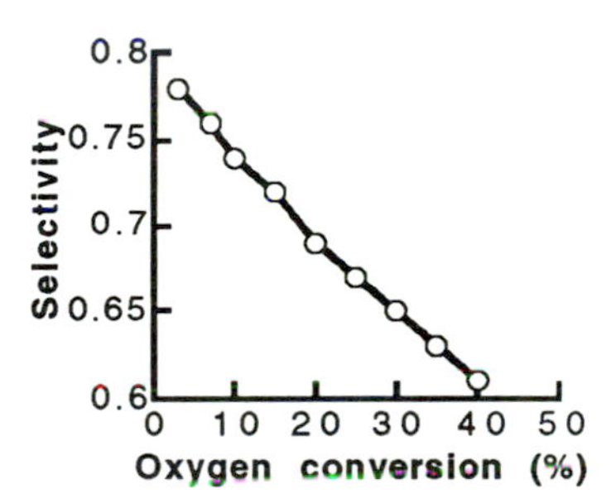

Fig. 5.8 The dependence of selectivity upon conversion for ethylene epoxidation over a Ag catalyst.

5.3 Forces on surfaces: non-ideality

In the treatments above a major assumption is made to simplify the derivations. For instance, in the Langmuir isotherm (Eqn 5.9) the terms in θ are separated on one side of the equation in order to linearise the equation. However, it is rarely the case that such plots are linear over a significant range of coverage due to a variety of factors, most importantly, lateral forces between atoms and molecules on surfaces. These result in the heat of adsorption for any species being strongly coverage dependent (see Fig. 5.9), usually decreasing as the coverage increases. This in turn means that the adsorption equilibrium constant is actually coverage dependent, and so the linearised form given above (Eqn. 5.9) generally cannot be expected to give a straight line.

There are many examples of these types of interactions, and perhaps their clearest manifestation is in the tendency of most adsorbates to order on the surface, that is, to arrange themselves in a regular, rather than random, manner. In this respect the discipline of surface science has proved invaluable. For brevity here we will give one example of this, namely the molecular adsorption of CO on Pd(111) where several structures on the surface can be observed by LEED (Section 2.1) due to the formation of ordered arrays of CO which repel each other into the minimum energy configuration (Fig. 5.10a). This is reflected in a decreased heat of adsorption as the coverage is increased (Fig. 5.10b). This shows that the rate constants for adsorption and desorption are strongly coverage dependent and, in the case of Fig. 5.9 here, these could change by approximately six orders of magnitude in changing the coverage from 0.3 to 0.5 monolayers. Clearly, then, gross assumptions are being made in the derivations described above and the rates of adsorption, desorption and reaction may be expected to show more complex behaviour than expected.

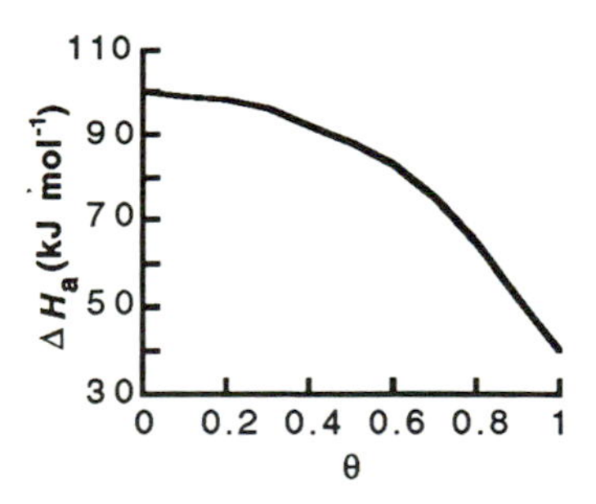

Fig. 5.9 The dependence of heat of adsorption upon coverage of the adsorbate.

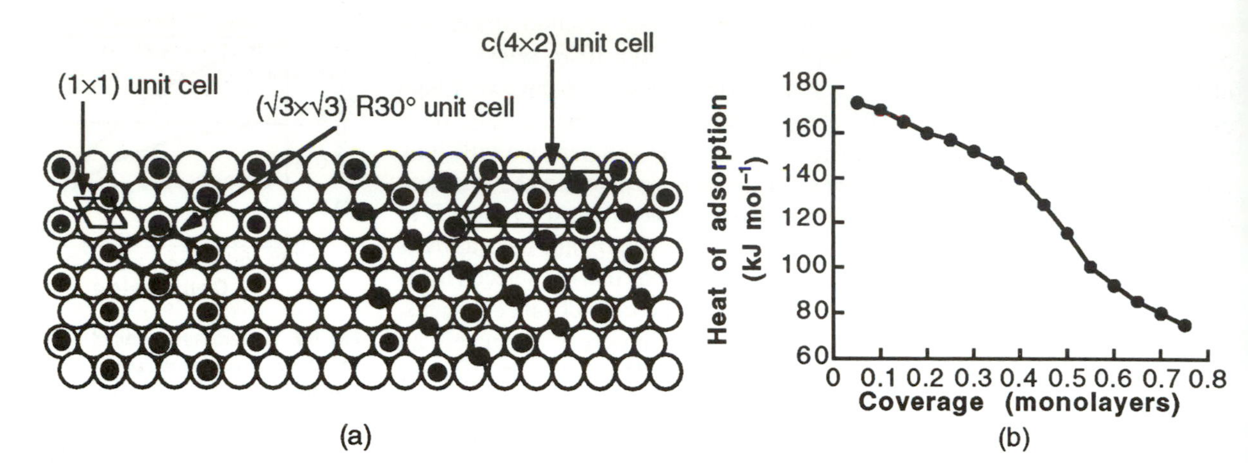

Fig. 5.10 (a) structures of CO adsorbed on Pt(111) at coverages of 0.33 and 0.5 monolayers and, (b) the dependence of heat of adsorption of CO upon coverage on the Pt(111) surface at 300 K. [Courtesy of Y.Y. Yeo, L Vattuone and D.A. King, J. Chem. Phys. 106 (1997) 392.]

5.4 Weakly held states and surface diffusion

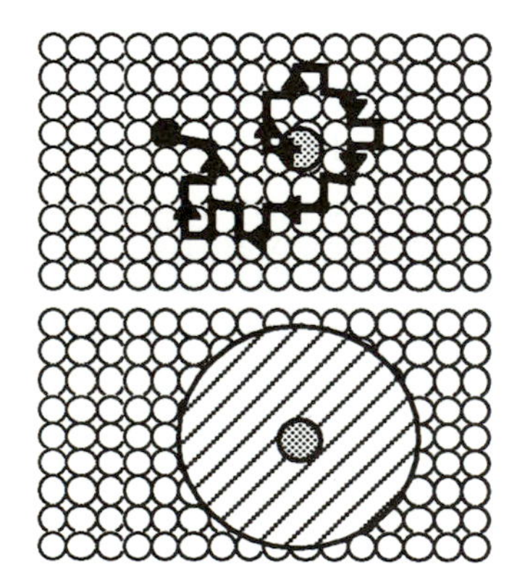

Fig. 5.11 Models of the diffusion of a weakly held molecule across a surface. Although the lifetime of such a species might be short, it may diffuse long distances in that time. The upper panel shows individual diffusional hops. The lower panel shows a 'diffusion circle' representing an average cross sectional area of the surface visited during the lifetime of the weakly held state. The number of surface sites visited is considerable.

In the section above, the effect of lateral interactions in the adlayer were shown to result in deviations from the ideal Langmuir behaviour of adsorbates. Another feature which is a deviation from the ideal is the presence of weakly held intermediate states in the adsorption processes. In the 'adsorption' section above it was considered that a molecule arrives at a surface *direct* from the gas phase and either finds a free site or does not, in which case it reflects back into the gas phase. However, as shown in Chapter 1 it is likely that most molecules first physisorb on a surface and then either adsorb molecularly or dissociatively into relatively strongly held, chemisorbed states. It is now clear that, although the lifetime in the physisorbed state is often low, the molecule can diffuse over long distances across the surface during its lifetime there (Fig. 5.11). The important points here are that: (i) the molecule can exist *over* filled sites in the chemisorbed layer; and (ii) such states can have a high probability of 'finding' an empty site on the surface, even though they may be low in coverage, during their short lifetime on the surface. In this book there is not space to consider the effects of such species on the detailed reaction kinetics, but the effects on adsorption alone will be briefly considered. The Langmuir adsorption relationship, Eqn 5.6, has to be modified as follows.

$$R_a = S_0.P.z\left(\frac{1-\theta}{1-\theta+K_p\theta}\right) \tag{5.33}$$

where K_p is the so-called precursor state parameter. If K_p is unity, then Eqn 5.33 reduces to the simple Langmuir form, and the precursor state has no effect on the kinetics. However, if it is low , then it has a substantial effect, as shown in Fig. 5.12, and the high diffusivity results in the adsorbate maintaining a high sticking coefficient at much higher coverages than would otherwise be the case. A reaction with such a process playing an important role is then mechanistically intermediate between the Eley–Rideal and Langmuir–Hinselwood models, though the kinetic rate equation can appear close to the former.

For a more detailed discussion of precursor state effects, see, A. Cassuto, and D.A. King, Surface Science 102 (1981) 388.

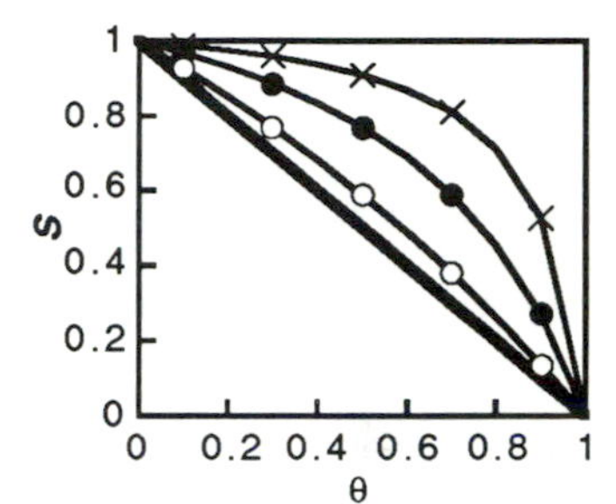

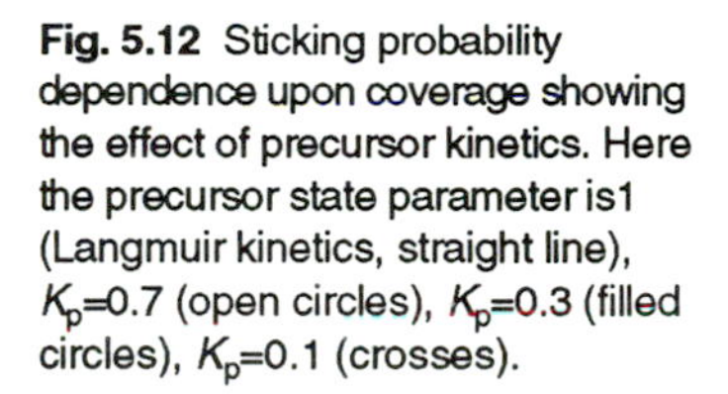
Fig. 5.12 Sticking probability dependence upon coverage showing the effect of precursor kinetics. Here the precursor state parameter is1 (Langmuir kinetics, straight line), K_p=0.7 (open circles), K_p=0.3 (filled circles), K_p=0.1 (crosses).

5.5 Surface dependence of reaction rates: the volcano principle

It has already been stated that the driving force for adsorption and catalysis is the surface free energy, but this driving force can be too strong to favour steady-state catalytic turnover. Thus Fig. 5.13 shows the catalytic reaction rate as a function of position of the metal in the transition series. This is a general, hypothetical plot, but illustrates the trends in reactivity. Metals in group 6 generally have very low catalytic turnover for any particular reaction. This is because the surface energy is so high that the metal concerned bind adsorbates very strongly, so strongly that it is difficult to break these bonds again and release products into the gas phase. Thus, in a CO/H_2 mix W tends to form carbides and oxides and becomes passivated. Such metals are self-poisoning. At the other extreme, the group 11 metals, the binding of adsorbates is very weak and so the coverage of the adsorbate is very low. Since the rate equation may be expected to be of the form shown in Eqn 5.17, then the rate will be low. An optimum exists in the intermediate metals where they bind molecules, but not too strongly, such that the product $\theta_A\theta_B$ is high, and there are free sites available for adsorption.

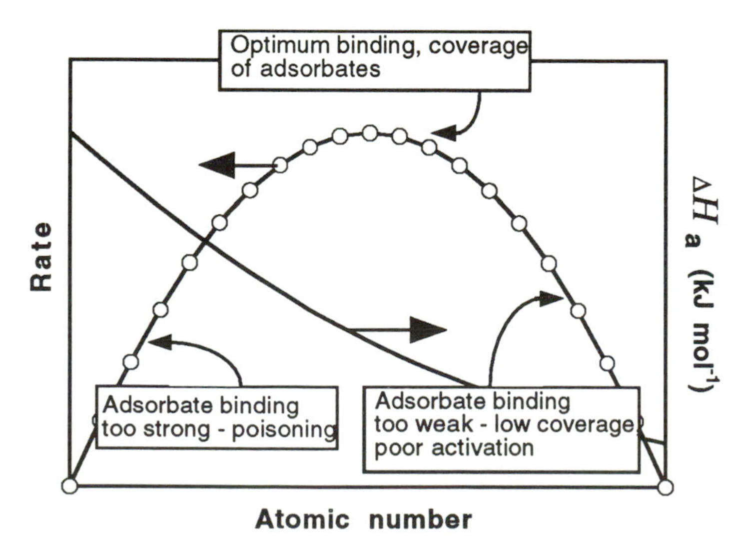

Fig. 5.13 A generalised 'volcano' plot for catalytic activity of transition metals.

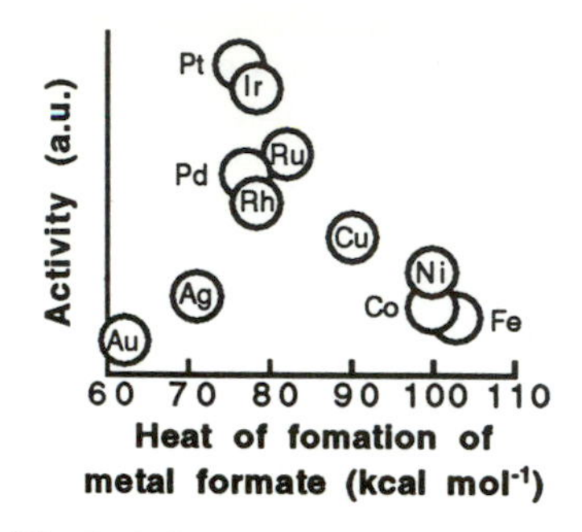

Fig. 5.14 A real example of data for the decomposition of formic acid on a range of transition metal surfaces. [Adapted from W. Sachtler and J. Fahrenfort, Proc. 2nd Intern. Congress on Catalysis, Paris (1961) 831.]

This kind of variation of catalytic activity is often seen, though not usually as neatly as shown in Fig. 5.13 because other factors, especially surface structure, are also of importance. An example of real data is shown in Fig. 5.14 for formic acid decomposition. Thus, changing the binding of the adsorbates or intermediates in a reaction will move the position of a metal around the 'volcano' maximum, and the addition of poisons, and promoters, or changing the surface structure composition can substantially affect the reaction rate. It may be the case, however, that one particular reaction (consider CO hydrogenation) may show optimum rates at different positions for different products. For example, if we compare the relative rates of methanation and methanol synthesis, we might expect curves of the kind shown in Fig. 5.15, with the optimum for methanol being to the right of the optimum for methanation. This is because the latter requires metals which adsorb CO weakly and do not break the CO bond, whereas methanation requires CO bond breakage and metals which can achieve that, without binding the resulting C and O too strongly.

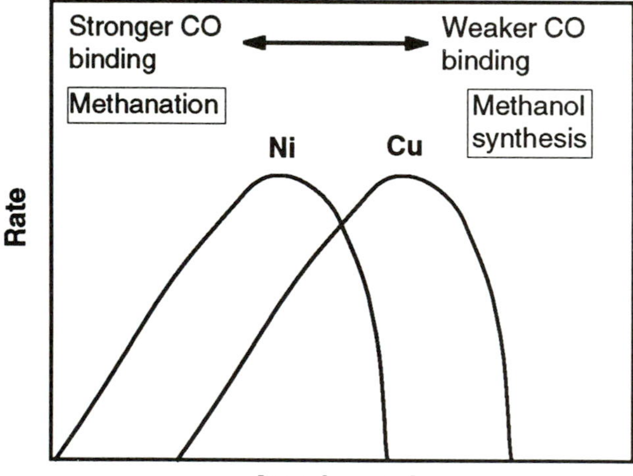

Fig. 5.15 Hypothetical 'volcano' plots for methanol synthesis and methanation to show the effect of the nature of surface bonding upon the profile.

6 Measurement of catalytic properties

6.1 Measurement of activity: microreactors to plants

The most accurate measure of the productivity of a catalyst is an industrial plant, for several reasons. First, it generally uses very large amounts of reactant and makes large amounts of product which can therefore be accurately measured and, secondly, the profit of a plant is drastically affected by a 1% change in activity and so the performance is accurately monitored.

However, industrial plants are inflexible, work in a very narrow window of temperature, pressure and flow rate, and the catalyst cannot be readily changed. For these reasons scaled down versions of reactors are used for catalyst testing as shown in Fig. 6.1, and the smallest of these is usually referred to as a microreactor.

There are a wide range of industrial reactors for heterogeneous reactions which can be broadly classified as follows.

1. **Fixed bed reactors**. Of these there are several types (Fig. 6.2).

   (a) Single pass. Reactant gas passes over the catalyst bed, products are collected at the end of the bed. For high conversion reactions. Catalysts are usually in pellet form.

   (b) Recycle. For low conversion reactions. Unreacted gas is recycled back to the front of reactor after separation of product, and 'make-up' gas added to maintain front end flow.

2. **Multitubular reactors**. Where extremely good heat conduction from/to the bed is required. Consists of many small diameter tubes in a bundle, each packed with catalyst and surrounded by coolant (e.g. in ethylene epoxidation) or more widely spaced tubes with heating gas (steam reforming).

3. **Fluidised bed reactors**. Give good mixing and good heat transfer between reaction medium and walls. For high conversion, high exothermicity reactions and for reactions which are mass transfer (diffusion) limited. The gas is fed at high velocity vertically through the bed of fine particles (often microspheres) and the bed becomes a turbulent fluid once a critical gas velocity is reached.

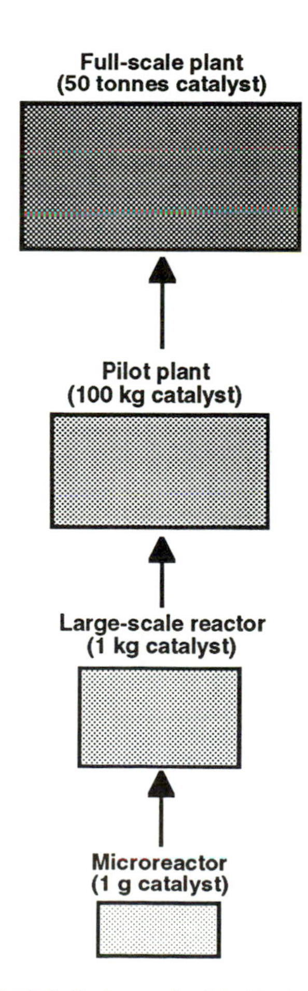

Fig. 6.1 Scale-up of catalyst testing.

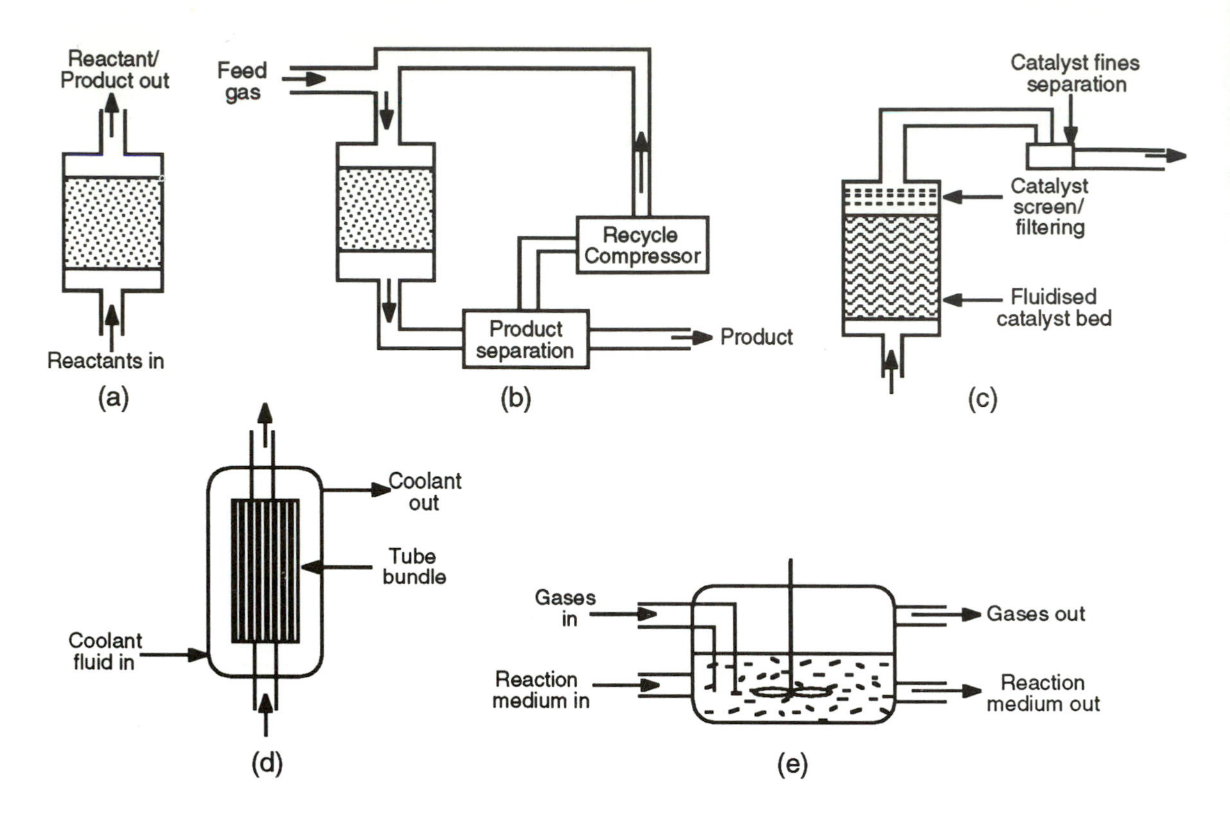

Fig. 6.2 Simplified schematic diagrams of various types of reactor (a) single pass (b) recycle (c) fluidised bed (d) multitubular (e) batch.

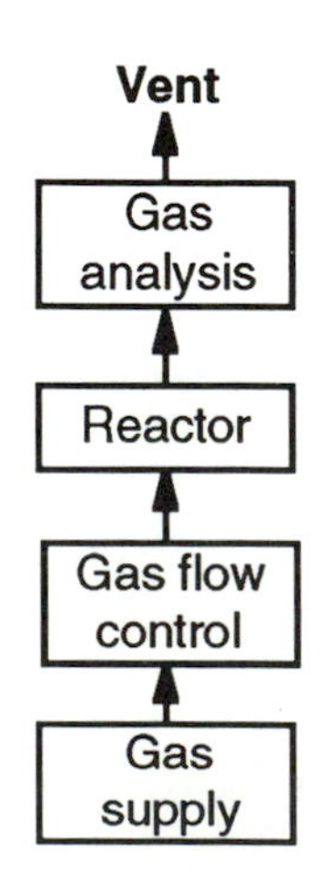

Fig. 6.3 General layout of a small-scale, continuous-flow reactor.

4. Batch reactors. Basically consists of a stirred pot into which the reactants and catalyst are placed, usually in the liquid phase, and products are separated after a certain length of time when high conversion has been achieved. Many low volume processes use such reactors, including the production of many pharmaceutical products. A number of variants on the theme exist, including recycle systems with product separation.

5. Flowing bed reactors. Some new reactors have the facility to recycle the catalyst from the reactor exit to the entrance, via a second unit which changes the surface state of the catalyst – usually by cleaning off carbon which poisons the reactivity of the catalyst, using a flow of oxygen. Such reactors are used, for instance, in FCC (fluidised catalytic cracking).

Smaller scale reactors of all these types can be made, but more commonly in academia and industry the standard form of microreactor is a small, fixed bed, taking ~1 g of catalyst, and the general layout used is shown in Fig. 6.3. Almost all the steady-state microreactor measurements for heterogeneous catalysis reported in the literature use just such a reactor, or variations on the theme.

It is important to distinguish two modes of reactor running, namely 'integral' and 'differential' mode. If a reaction is run to produce optimum conversion, then there is an increase in products down the length of the catalyst bed, and a decrease in concentration of the reactant. This makes it almost impossible to measure the kinetics of the process, since if we imagine the simplest reaction kinetics, such as the Langmuir equation (see Chapter 5), the rate varies along the bed, because the reactant concentration varies; this is illustrated in Fig. 6.4b. A reactor run in this mode is called an integral reactor, while one with essentially constant amounts of reactant, running at low conversion, is called a differential reactor (Fig. 6.4a). A tube can be run in either mode by varying the conversion. At low conversion a tubular reactor is a differential reactor, but an integral one at high conversion. In contrast, a well-stirred reactor is always a differential reactor (that is, all the catalyst sees the same concentration of product and reactant), providing the reactant and catalyst are well mixed; it is then called a 'stirred-pot' or CSTR (continuously stirred tank reactor). One such is illustrated in Fig. 6.2e.

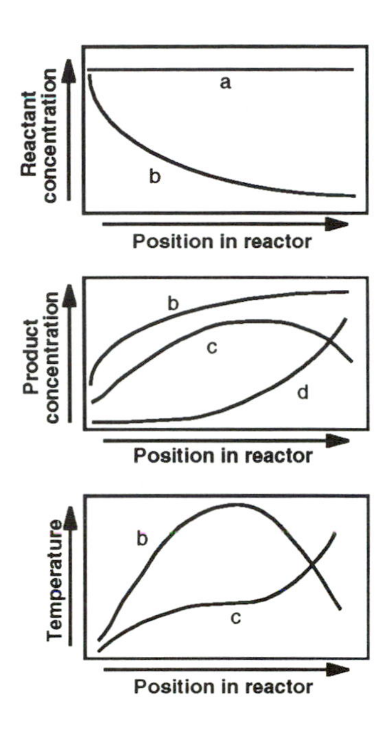

Fig. 6.4 Reactant, product and temperature variation down the bed of a reactor run in integral mode (b) and differential mode (a). In the products graph curves (c) and (d) represent an intermediate and final product, for instance a partially oxidised product and a completely oxidised one. For the latter case the temperature profile may be like (c) in the bottom panel.

Engineering considerations for small scale reactors

Some points of engineering are essential to the design of catalytic reactors, but it would need at least a book to do this subject justice; here we will consider the most important points related to design. There are two main considerations – heat and mass transfer.

In order to extract kinetics it is important that the temperature of the reactor is kept constant and for any reactor run at high conversion this is difficult. Consider Fig. 6.4, which shows fast conversion at the front of the bed where the concentration of reactant is highest, and slower conversion further along the bed. The result is inhomogeneous heating for an exothermic reaction, and the profile of T along the length is strongly dependent on the nature of the reaction. In the simple $A \rightarrow B$ conversion, there is a maximum temperature at some point in the bed, the exact point depending on the heat flow characteristics of the reactor and the flow velocity (curves b). For a more complex reaction, for instance, a sequential reaction with burning of the product, the 'hot-spot' in the reactor may be at the exit. In order to minimise these variations in temperature down the bed we need: (i) a low conversion; and (ii) good heat transfer to the walls, which is best achieved by use of a small diameter reactor tube (~6 mm o.d. or less).

Mass transfer and diffusion limitations are also aided by small bed widths and by use of small catalyst particles. However, very small particles cannot be used because reactor blocking and large pressure drops across the bed then become a problem. If mass transfer is a problem then it can be alleviated to some degree by extra disturbance of the flow pattern in the bed by having a profiled, or baffled wall, though this itself can cause problems with dead spots in the reactor.

6.2 Physical structure of catalysts

As discussed earlier, the total surface area and pore volume are important properties of a catalyst that determine how active it is. Thus a range of methods have been evolved to measure these parameters and are outlined in brief below.

Measuring total surface area

In order to measure the surface area of any porous material a reliable, non-specific method is needed. Physisorption is fairly material-independent in the sense that N_2 molecules at low temperature tend to form a monolayer which depends only on the size of the N_2 molecule, which is a known standard (0.16 nm^2); the molecules pack together as closely as they can, independent of the substrate atomic structure. The equation for this approach was derived by Brunauer, Emmett and Teller (the BET equation), recognising that multilayers of physisorbed adsorbate can form as the monolayer is being filled. The equation is presented in linearised form as follows

$$\frac{P}{V(P_0-P)} = \frac{1}{V_m.C} + \frac{(C-1)}{V_m.C} \cdot \frac{P}{P_0} \tag{6.1}$$

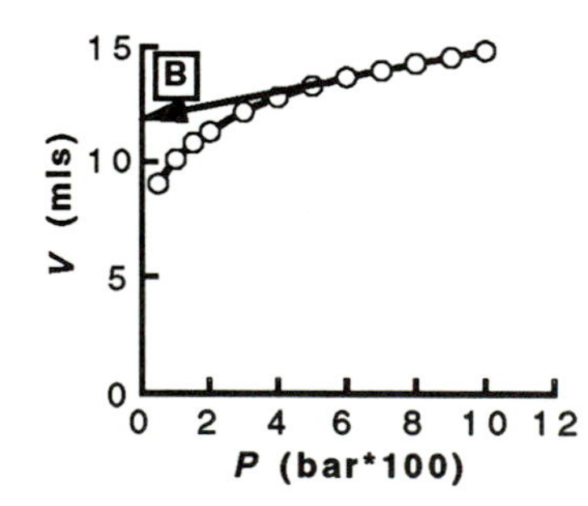

Fig. 6.5 Typical form of an isotherm of uptake dependence upon equilibrium pressure.

where P is the equilibrium pressure for a particular surface coverage, represented by V, the volumetric uptake of nitrogen V_m is the volume required to cover the surface to one monolayer thickness and C is a constant.

The experiment is carried out by measuring the amount of N_2 adsorbed at 77 K on a sample as a function of N_2 pressure over the sample – the higher the pressure, the higher the amount adsorbed. The form of such a curve is shown in Fig. 6.5 and the linearised form is shown in Fig. 6.6. The surface area is determined from V_m, the monolayer volume, which in turn is extracted by measuring the intercept and slope of Fig. 6.6. In this particular case they are as follows.

Intercept = 0.00032 $mls^{-1} = (V_m.C)^{-1}$

Slope = 0.072 $mls^{-1}.\ bar^{-1} = \dfrac{(C-1)}{V_m.C.P_0}$

and so $V_m = 13.8\times10^{-6}\ m^3$

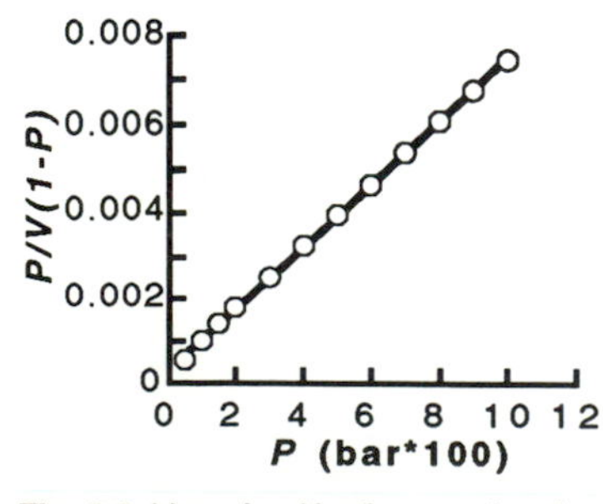

Fig. 6.6 Linearised isotherm using the BET relationship.

and this is the volume of gas (at STP) which is needed to form the monolayer of adsorbate. Thus the area of the surface involved is given by

$$SA(m^2) = V_m(m^3) \times \frac{N_A}{V_A}(\text{molecules}/m^3) \times A \tag{6.2}$$

where A is the area of the nitrogen molecule and N_A/V_A is Avogadro's number per unit volume of gas. The specific surface area (SSA) is then this number divided by the weight of catalyst used in the measurement. For the example above.

$$SSA = \frac{SA}{W} = \frac{60\ \text{m}^2}{4.7\ \text{g}} = 12.8\ \text{m}^2\ \text{g}^{-1} \tag{6.3}$$

The experiment to measure the monolayer volume can be done in several ways.

1. Point B method. Here a single point measurement is often made, as at a single value of N_2 partial pressure (usually for $P/P_0 \sim 0.1$), and the volumetric uptake is taken as the measure of V_m, or simple extrapolation of a set of data is taken from the 'plateau' in the isotherm (shown in Fig. 6.5). Since the detailed shape of such isotherms is strongly dependent on the pore size distribution (below) then the result can be in error. However, in practice, for microporous and mesoporous materials, the values determined are often within 10% of the surface area.

2. Volumetric method. Here the sample is maintained in a vacuum and aliquots of N_2 gas are allowed into the fixed volume containing the sample, held at 77 K. After each aliquot of gas, the sample adsorbs the N_2 and slowly evolves to an equilibrium surface coverage/gas pressure. Determination of the latter requires an accurate pressure gauge. The former is determined from the known amount of gas let in, and the shortfall of the equilibrium pressure from that expected if no adsorbate were there. Since an accurate isotherm is required this is usually a very lengthy experiment needing many hours to reach true equilibrium at each point.

3. Dynamic method. In this way a complementary measurement of uptake and BET methods can be carried out conveniently and quickly in a single experiment. A flow of an inert, low molecular weight carrier gas (such as He) is maintained over the sample maintained at 77 K in a small tubular reactor. Then N_2 is switched into the flow, at the level of approximately 10% by volume. As shown in Fig. 6.7, the mass spectrometer detector shows no presence of N_2 in the exit stream until the monolayer is saturated, when the N_2 'breaks through' and is measured. If the N_2 is then switched out a desorption isotherm is obtained and this trailing edge can be analysed by integration to obtain P, V, relationships, for use in a BET plot. Such methods often give remarkably good agreement for the surface area (within 10% of each other). This approach is much faster than that of method 2 (~20 mins as opposed to several hours).

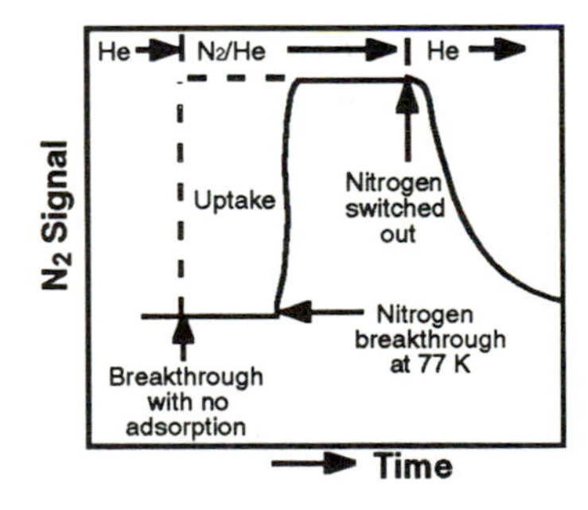

Fig. 6.7 Dynamic method for surface area determination, using nitrogen adsorption at 77 K.

Pore volume determination

From a wider range plot of nitrogen uptake to higher pressures than in Fig. 6.5, information about the distribution of pore volumes and therefore pore sizes can be obtained. This is because the filling of pores with condensed nitrogen beyond the monolayer point is related to the radius of the pores and the gas pressure; the smaller the pore radius the more easily it is filled with condensed nitrogen. The relationship between these properties is given by the Kelvin equation.

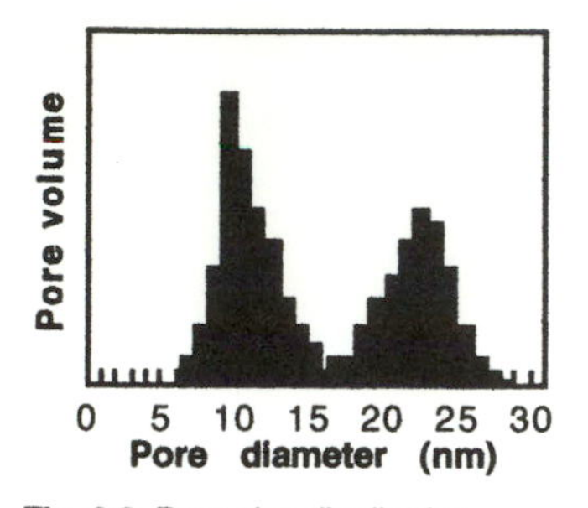

Fig. 6.8 Pore size distribution, showing a bimodal distribution of the pores.

$$\ln\left(\frac{P}{P_0}\right) = \frac{2\sigma V_0}{r_K RT} \quad (6.4)$$

where V_0 is the molar volume, σ is the liquid surface tension and r_K is the 'Kelvin pore radius'. The distribution of pore sizes is then as follows

$$\Delta S = \frac{2\Delta V_P}{r_K} \quad (6.5)$$

where ΔS is the differential surface area associated with a particular average pore radius and ΔV_P is the pore volume determined over a narrow range from the *P–V* isotherm. An example of such a determination is shown in Fig. 6.8. For a microporous solid such as a zeolite, almost all the area will be associated with the well-defined small pores, while for most catalytic materials the major part of the area will be in mesopores.

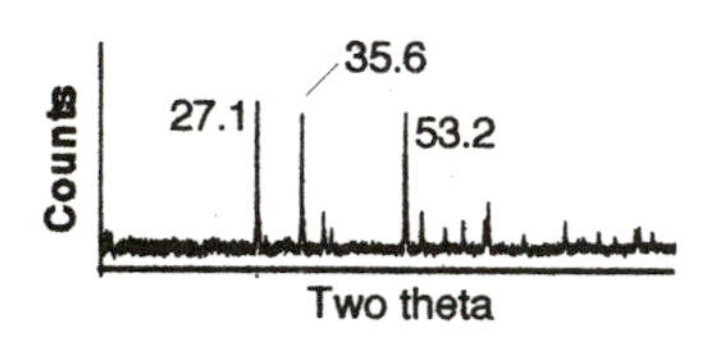

Fig. 6.9 Diffraction pattern from a rutile material, $FeSbO_4$. (Courtesy of the Catalysis Research Centre, University of Reading).

Bulk structure – X-ray diffraction (XRD)

The classical technique for the determination of the bulk structure of crystalline solids is XRD and an example diffraction pattern for an $FeSbO_4$ ammoxidation catalyst, made with an Fe:Sb atomic ratio of 0.5, is shown in Fig. 6.9. The peaks are associated with diffraction from particular planes in the lattice and are dictated by the Bragg equation

$$n\lambda = 2d\sin\theta \quad (6.6)$$

where n is an integer, λ is the X-ray wavelength, d is the particular lattice plane spacing and θ is the Bragg diffraction angle (half the deviation of the diffracted beam from the incident beam). Since catalysts are generally powdered materials, XRD has to be carried out by analysis of diffraction circles, rather than spots, because crystallite orientation around the X-ray incidence angle is random.

Some limited information about the particle sizes in solids can be obtained because, when the crystallites of the catalyst are small, the diffraction lines are broadened by the limited range of order; indeed, if the particles are too small then the lines are too diffuse to measure. The average particle size can be determined by application of the Debye–Scherrer equation

$$B = \frac{0.893\lambda}{d\cos\theta} \quad (6.7)$$

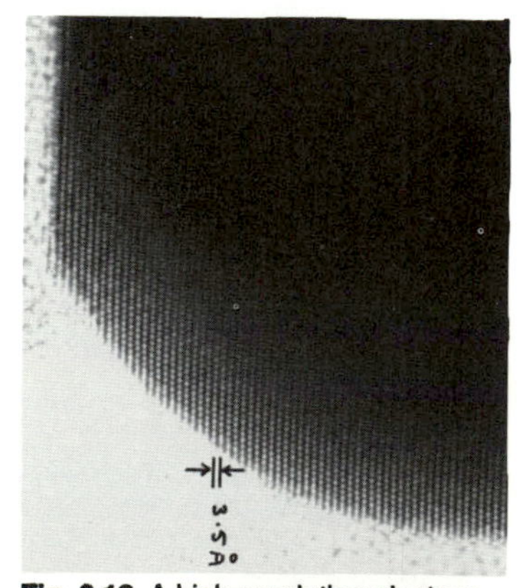

Fig. 6.10 A high-resolution electron microscopy image of a single particle of the $FeSbO_4$ whose XRD pattern is shown in Fig. 6.9. Individual atomic positions can be identified. [Fig. 6.10 and 6.11 Courtesy of M. Goringe and E. Bithell, Oxford University and from Bithell et al., Physica Status Solidii 46 (1994) 461.]

where B is the broadening of the diffraction peak, λ is the X-ray wavelength (0.154 nm for Cu-K_α, which is usually used), θ is the angle of the diffraction peak, d is the crystallite size.

Electron microscopy and particle size distributions

Electron microscopy is useful for obtaining structural information about catalysts, but the technique exists in a variety of forms, some of which can spatially resolve to the atomic level (see Fig. 6.10) and most can now carry out elemental analysis by, for instance, EDAX (energy dispersive analysis of X-rays) (see Fig. 6.11). Often, qualitative information is obtained, such as that shown in Fig. 1.2, but detailed computer analysis of such images can be made to determine particle size distributions, including the metal particles on a supported catalyst (Fig. 6.12). Care has to be taken with such data however, since particle sizes determined by electron microscopy often turn out to be bigger than those determined on the basis of other techniques such as surface area measurements. This can be because of an inadequate determination of the distribution of particle sizes or the fact that some particles may be composites of smaller units or have some porosity themselves.

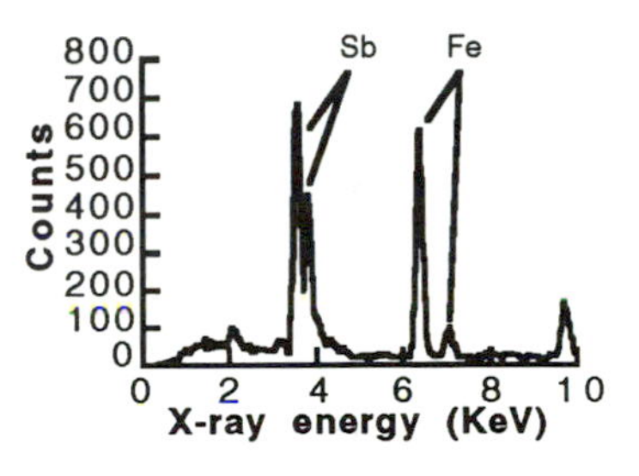

Fig. 6.11 EDAX analysis of $FeSbO_4$.

EXAFS

Synchrotron radiation is now used widely because synchrotrons produce a wide range of wavelengths of light with a high flux of radiation. Such a facility can be used, for instance, to carry out X-ray diffraction measurements. A technique for the analysis of materials which has been developed over recent years is EXAFS (extended X-ray absorption fine structure) and this has been widely applied to catalysis. It is a bulk, averaging technique, but because of the local nature of the interaction which produces the effect it can be applied to understand the structure of very small particles.

The principle of the technique is highlighted in Fig. 6.13a, and relies on the interference between an outgoing photoelectron wave (the photoelectron produced by photoionisation of the emitter atom) and the backscattering wave off adjacent nuclei. This causes energy-dependent fluctuations of the measured absorption of the solid due to destructive and constructive interference (Fig. 6.13b). These fluctuations have a strong dependence on the distance between the two atoms involved, and on the number of surrounding atoms. In fact, these can be approximately represented as follows, where $X(k)$ is the deviation of the scattering from a smooth background absorption.

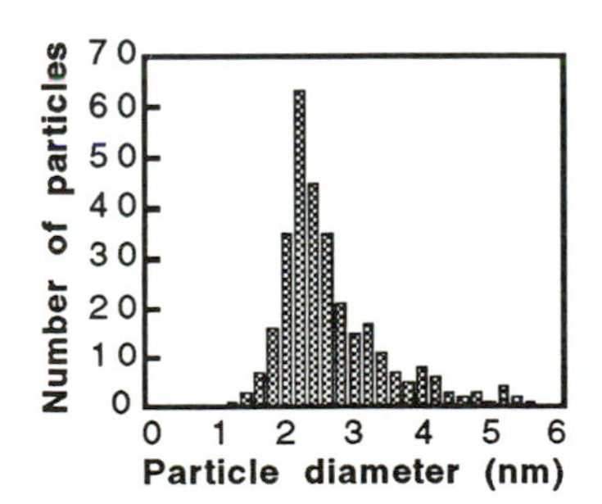

Fig. 6.12 A histogram showing the metal particle size distribution for a supported catalyst.

$$X(k) \propto -\ \Sigma_i \frac{N_i}{kR_i^2} \sin[2kR_i + \phi_i(k)] \qquad (6.8)$$

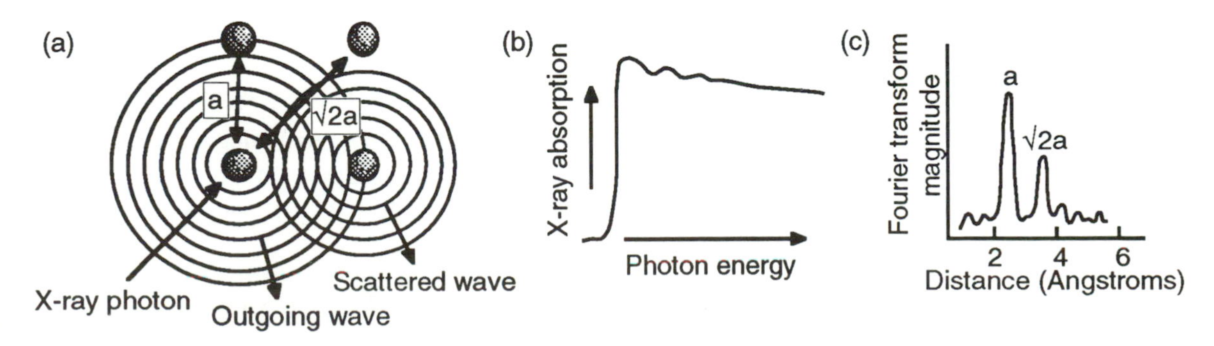

Fig. 6.13 (a) Showing the basic principle of photoelectron scattering in EXAFS. (b) The form of EXAFS raw data showing the absorption edge sinusoidal 'wiggles'. (c) The type of radial distribution plots extracted from EXAFS data.

Here k is the photoelectron wavevector (related to wavelength), N_i is the effective coordination number of an atom in the ith shell (the summation is over i shells of atoms surrounding the emitter), and R_i is the distance of the ith shell from the emitter. The sine term dictates the oscillations in the amplitude and their wavelength dependence, where ϕ_i is a phase shift term. Such data are subjected to Fourier analysis to extract the radial distribution of shells of neighbouring atoms, from which the structural relationship can be directly determined. Such a curve is shown in Fig. 6.13c, and shows two major neighbouring distances at the lattice parameter and $\sqrt{2}$ times that value.

Thus, in principle, the structure of a small particle can be determined, and some information can be obtained about its relationship to other atoms (for instance bonding to a support medium). Furthermore, catalysts with different atomic components (e.g. oxides or alloys) can be analysed for each separately. However, a limitation is that this is an averaging technique, giving information on average numbers of nearest neighbours and of average bond lengths for nearest neighbours, and so, for catalysts of low dispersion (which many are) only bulk information is obtained, since the bulk represents the vast majority of atoms measured.

6.3 Surface structure

Specific chemisorption

This technique is useful for determining the area of a particular active component of a catalyst, especially for metals, and can sometimes be used to determine the structure of the surface. In total area measurements, as described above, physisorption of nitrogen is used and is indiscriminate in adsorption site. To determine the active site contribution a specific method is used, which involves chemisorption of reactive molecules such as O_2, CO, H_2, N_2O. Examples of the application of such a technique in a dynamic flow system are given in Fig. 6.14. The method relies on an assumption of the surface structure (although densities of surface atoms do not vary very widely, being $\sim 1\times10^{19}$ m^{-2}) and on surface science determinations of the stoichiometry of adsorption (adsorbed atoms or molecules per surface metal atom).

An example calculation is as follows.
Total uptake of O_2 is 0.28 mls at STP for 3 g of a 10% Ag/Al_2O_3 catalyst (measured from the uptake seen in Fig. 6.14a). This is equivalent to $\frac{0.28}{22400}\times$ 6×10^{23} molecules = 7.5×10^{18} molecules of O_2. Assuming that the surface of the Ag is the close-packed one, with a density of 1.4×10^{19} molecules m^{-2}, and that the saturation density of oxygen is 0.5 monolayers in the atomic form (from literature values), then

$$\text{Area of Ag} = \frac{\text{molecules } O_2 \text{ adsorbed}}{\text{molecules per sq. metre}} = \frac{7.5\times10^{18}}{3.5\times10^{18}} = 2.1\ m^2$$

$$\therefore\ \text{metal area / g catalyst} = 0.7\ m^2\ g^{-1};\ \text{metal area /g metal} = 7\ m^2\ g^{-1}$$

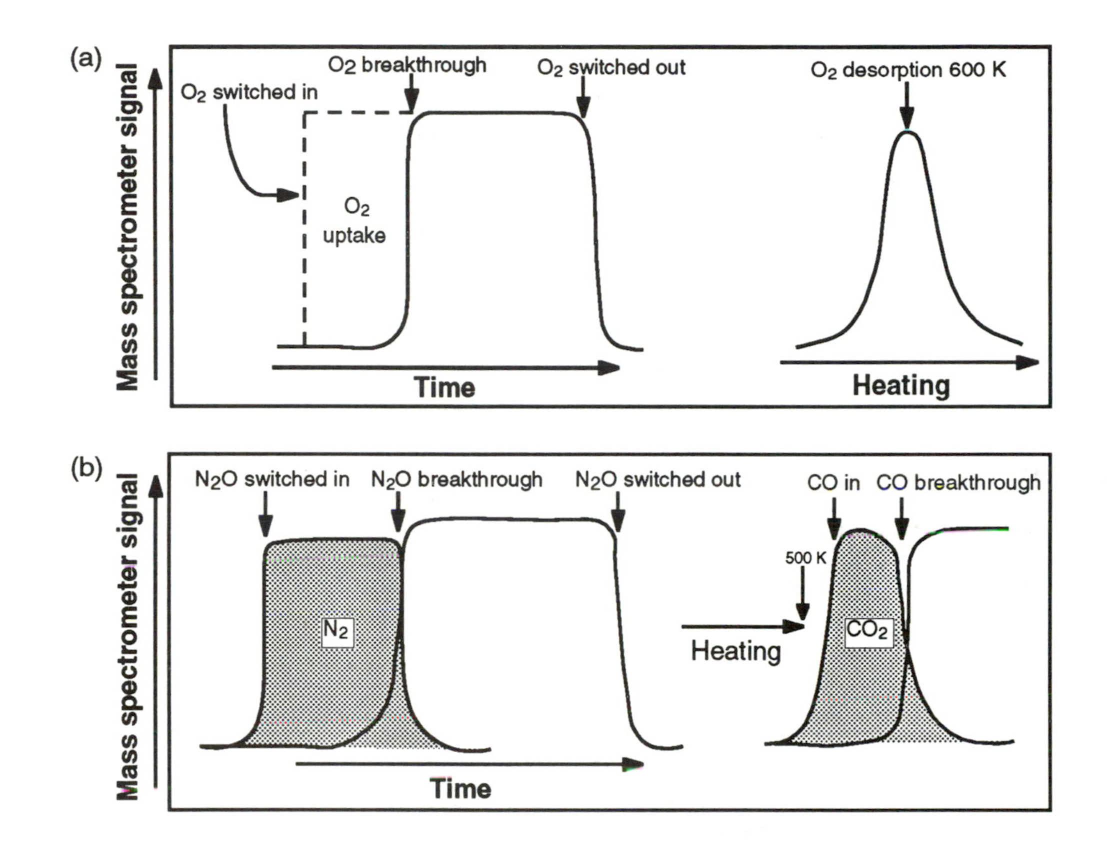

Fig. 6.14 A dynamic determination of metal surface areas: (a) oxygen adsorption on a Ag catalyst and (b) N_2O adsorption on a Cu catalyst.

Surface analysis

A wide range of techniques are available to analyse chemically the composition of a catalyst surface, and these are classified in Fig. 6.15. They include photon-in techniques such as XPS and IRAS (Infra-red absorption spectroscopy) and electron-in methods such as AES (Auger electron spectroscopy) and electron microscopy. Perhaps the most utilised technique for surface analysis nowadays is XPS, due to its surface sensitivity, and an example of its application was given earlier (Chapter 3). However, a further useful feature of XPS is its sensitivity to the valence state of atoms and to the environment of atoms in organic molecules. Thus, if a Bi_2MoO_6 catalyst used for ammoxidation is reduced *in situ* in an XPS machine, the data shown in Fig. 6.16. are obtained. This directly indicates reduction of some Bi^{3+} to Bi^0, while reduction of some Mo^{6+} to Mo^{4+} and Mo^{5+} has also occurred (not shown). This is a very important technique for assessing changes in a catalyst before and after reaction. Furthermore, the state of adsorbates themselves can be monitored and, for example, different C atoms in organic molecules can be distinguished. Acetate adsorbed on a metal surface manifests two well-separated C(1s) peaks due to carboxylate and methyl carbons. The

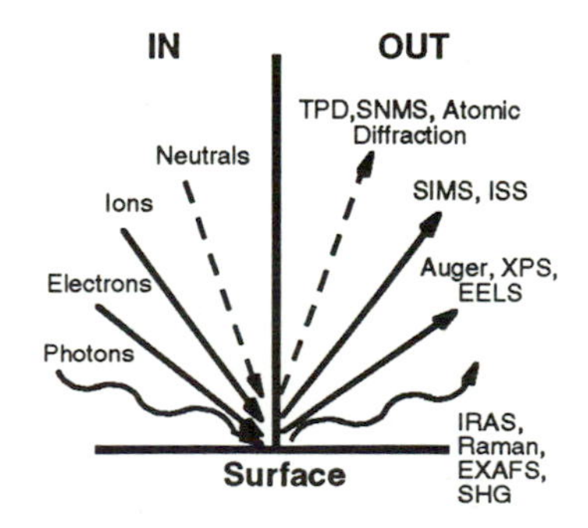

Fig. 6.15 A diagram of some of the techniques now used for surface analysis.

carboxylate is shifted to higher binding energy due to removal of electron density by the adjacent oxygens (Fig. 6.17). The O(1s) shows a single, narrow line due to identically situated oxygen atoms – both binding to the surface.

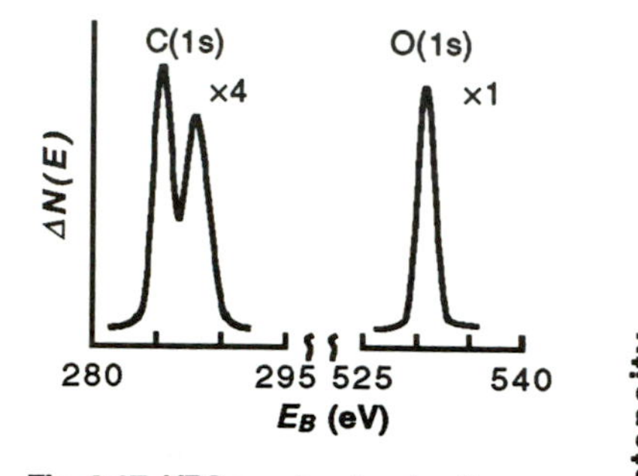

Fig. 6.17 XPS spectra showing the presence of acetate on a Cu surface. The two carbons in the molecule are distinguished while the two oxygen bind identically to the surface. [Adapted from M. Bowker and R.J. Madix, Appl. of Surf. Sci. 8 (1981) 299.]

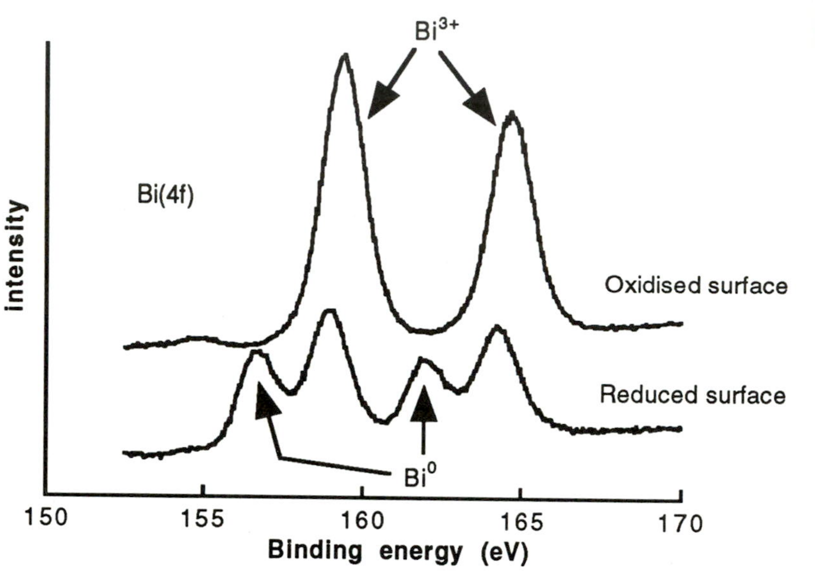

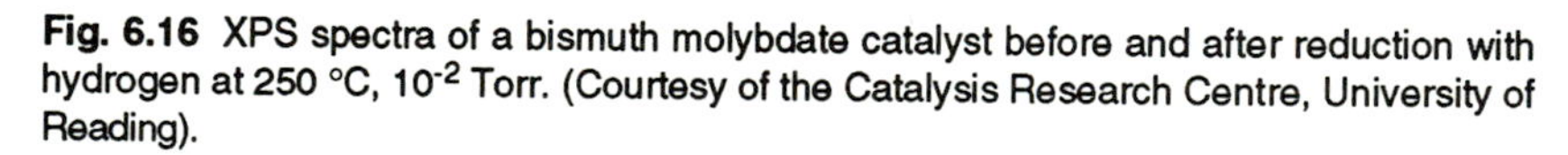
Fig. 6.16 XPS spectra of a bismuth molybdate catalyst before and after reduction with hydrogen at 250 °C, 10^{-2} Torr. (Courtesy of the Catalysis Research Centre, University of Reading).

III Applications: catalysis for the benefit of humans

Catalysis is a beneficial technology for a wide variety of reasons and these can be summarized as follows.

1. **Environmental efficiency**. Catalysts generally enable chemical conversions to take place at lower temperatures, and under milder conditions than would otherwise be necessary. As a result the reactions are also normally less polluting because careful control of the emissions (and their removal with clean-up catalysts) results in less wastage of raw materials and is less costly.

2. **Economics**. Most countries depend on the petrochemical industry as the basis for their modern, materials-based economies. Most materials are produced by the chemical industry, ranging from plastics to pharmaceuticals, and approximately 90% of all these processes use catalysts at some stage. Without catalysis the nature of modern economics would be completely changed. Life itself is dependent on the use of natural catalysts (enzymes) for cell function and so, without these, life would not have evolved.

Thus, catalysis is an essential phenomenon and in this section we will review some of the main processes to which catalysis is applied for the material benefit of humans.

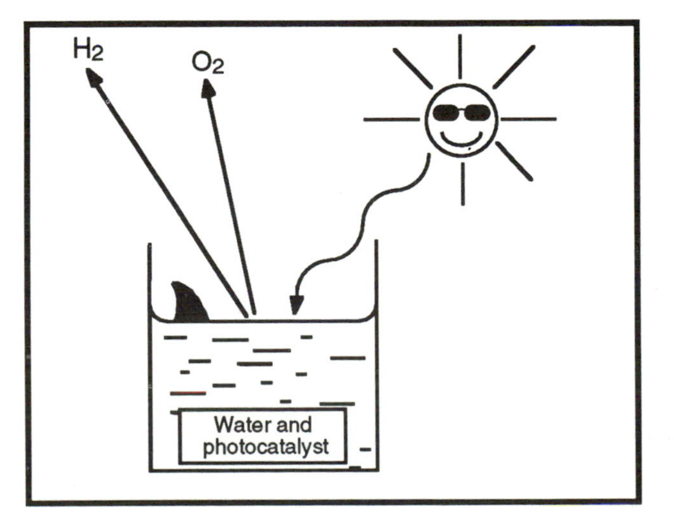

7 Raw materials and their conversion

7.1 Types of feedstock

The major volume sources of chemicals are coal, oil and natural gas (Fig. 7.1). The dominant source for the chemicals industry in the early part of the century was coal which was converted to a variety of products by high temperature reaction, and which was used for both domestic and industrial use. The products included coke, town gas for fuel and coal tar for chemicals production.

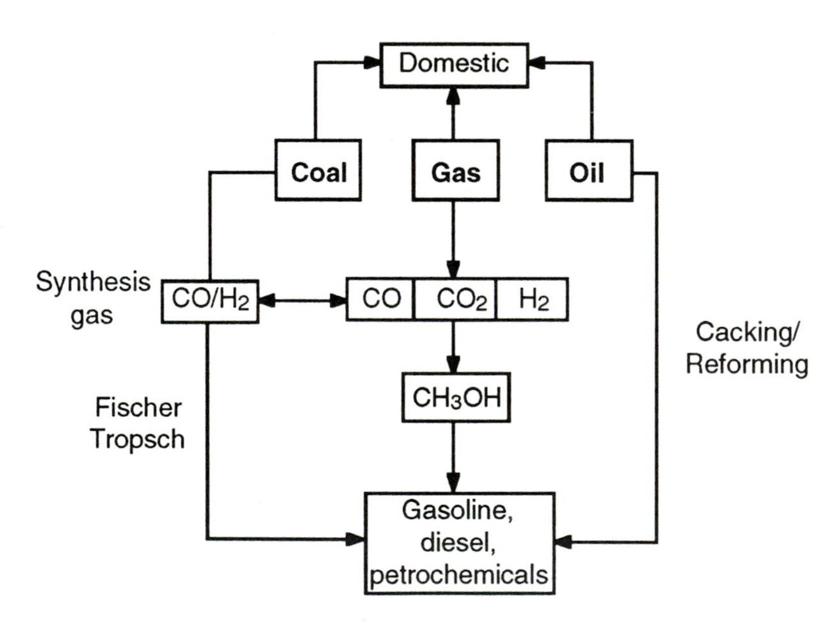

Fig. 7.1 The domestic and industrial use of fossil fuels.

The use of coal in the chemical industry was phased out after World War II and replaced with increasing supplies of oil, especially from the Middle East where large reserves were discovered. Oil is converted to a variety of products by cracking, as described in more detail below, ranging from bitumen for use on roads, to petrol for cars, to ethene for plastics production. It is presently the major source of carbon for chemicals production and fuel.

Since the 1960s there has been increasing use of natural gas in the chemical and fuel sector of the economy, especially in the production of some of the high volume chemical products such as ammonia and methanol. In the former case the hydrogen is extracted from the methane to make ammonia, the carbon is not used and represents a large source of CO_2 to the atmosphere (one plant produces approximately 200,000 tonnes of CO_2 per year). Use of all of these fossil feedstocks results in most of the carbon used being ultimately emitted as CO_2 into the atmosphere. Some of the solid products (plastics for instance) may end up in landfills and may be converted to methane. Both CO_2 and methane are severe greenhouse gases and evidence is mounting that strong legislation will have to be introduced to reduce annual emissions of this kind. This can be achieved, to some degree, by efficient recycling (of plastics for instance), but is almost certainly going to require new, pollution-free processes, and this will be discussed in more detail in Chapter 10.

7.2 Feedstock conversion

Figure 7.2 shows the crude pattern of feedstock conversion and products formed. If we begin with the main carbon source for the petrochemical industry, crude oil, then the first process is separation of the fractions of light molecules, intermediates (naphtha) and heavier fractions the latter being of a wide range of molecular weights with carbon numbers greater than 12. This is achieved by fractional distillation, the higher molecular weight species having higher boiling points. The range of boiling temperatures is from ambient temperature, for the light gases, to above 500 °C, for bitumen tars.

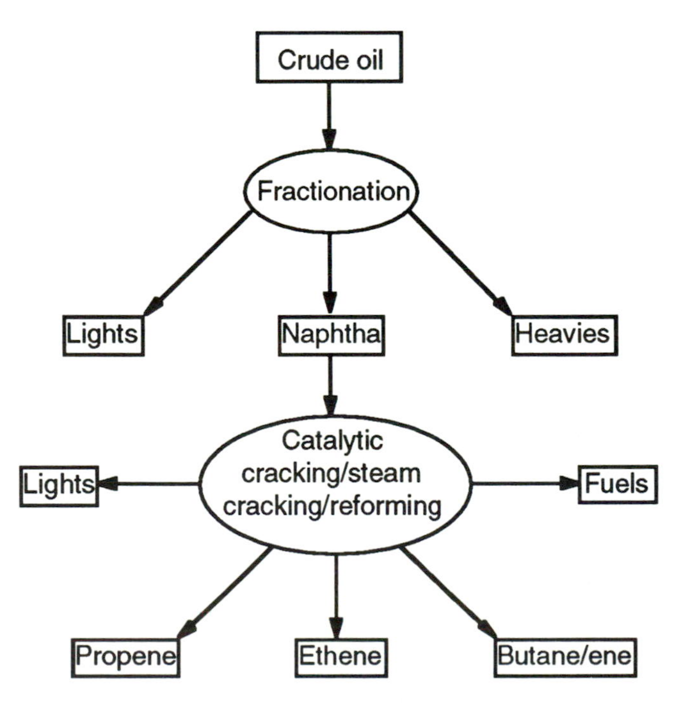

Fig. 7.2 Simplified schematic diagram of the conversion of crude oil into some of its major usable products.

The most useful molecules for chemical synthesis and for fuel are produced by catalytic cracking (cat cracking and hydrocracking) of the higher chain length alkenes, to produce molecules closer to C_8 in an unsaturated form. The major reason this is so important is the huge global demand for fuel for vehicles. Some of the molecules, particularly cyclic paraffins, are reformed, again to produce functionality by hydrogen loss and also to open up the rings. This produces fuels with high 'octane' rating which is needed to give fuel the 'anti-knock' characteristics desirable for use in modern engines. The catalysts involved in these various processes are listed in Table 7.1. The most useful molecules for chemical synthesis are the short chain alkenes ethene and propene, which are produced in huge quantities from the crackers, though mainly from non-catalytic steam cracking of naphtha.

Table 7.1. Catalysts used in oil conversion.

Process	Catalyst
Hydrodesulph-urisation	CO/Mo sulphides/γ-Al_2O_3
Catalytic cracking	Al_2O_3, zeolites
Hydrocracking	Pd/zeolite
Reforming	Pt (+ additives such as Sn)/η-Al_2O_3

Natural gas is now readily available in many parts of the world and so has become widely used in simple molecule synthesis since the 1960s. It is mainly used in ammonia and methanol synthesis, as described in the next section. Methane is a very stable hydrocarbon (hence its abundance in the earth) and so needs rigorous conditions for its reaction and conversion. Currently it is converted by steam reforming, by oxidation or by the use of both processes.

In steam reforming the gas is heated to high temperature and passes through a tubular reactor containing a catalyst consisting of Ni supported on alumina and is heated to ~1000 K by burning gas on the outside of the tubes. This results in oxidation of the methane by water and thus production of hydrogen.

$$CH_4 + H_2O \rightarrow CO + 3H_2 \tag{7.1}$$

The reaction is difficult to complete in one step and often requires a secondary reformer to oxidise the remaining methane (from ~10% to <1%).

$$CH_4 + O_2 \rightarrow CO + 2H_2 \tag{7.2}$$

Other reactions occur during this catalysis to form CO_2 as a major product, such as the water gas shift reaction shown below (Eqn 7.3). This latter type of reaction is used to extract more H_2 from H_2O in a later catalytic reactor (running at low temperature, since it is an exothermic reaction) using other catalysts. The products of methane conversion are used in the ways outlined below (Section 7.3).

$$CO + H_2O \rightarrow CO_2 + H_2 \tag{7.3}$$

Coal was a major feedstock earlier in this century for the production of synthesis gas. Coal was anaerobically carbonised to produce mainly solid fuel (in the form of 'coke') and 'town' gas (mainly CO/H_2), together with small amounts of higher molecular weight synthetic chemical precursors. More recently it has mainly been used in times of emergency to produce fuels and other hydrocarbon products, when oil supplies were cut-off. This was notable in Nazi Germany, when oil could not easily be imported, and in South Africa during the oil embargo crisis of the Apartheid era. The South Africans built a huge plant at Sasolburg to convert coal to fuels. It used Fischer–Tropsch technology invented in Germany. The process involves coal gasification as the first step, that is, steam reforming of coal at very high temperatures.

$$2C + 2H_2O \rightarrow 2CO + 2H_2 \qquad (7.4)$$

Thus synthesis gas is obtained which can be used for a wide range of products as outlined below.

7.3 Products

From natural gas

As has been seen, the main way of using oil is by distillation, cracking and reforming to produce a wide range of products, while the main way of using coal and natural gas is by steam reforming to produce synthesis gas which can then be used in a variety of ways. A great deal of methane is used to make methanol, as shown in Fig. 7.3, and the catalyst for the synthesis is a very efficient one operating at low temperatures (~520 K) and producing methanol in high selectivity (~99%). This methanol, in turn, is converted to a wide variety of products, as shown in Fig. 7.4, many of them produced catalytically.

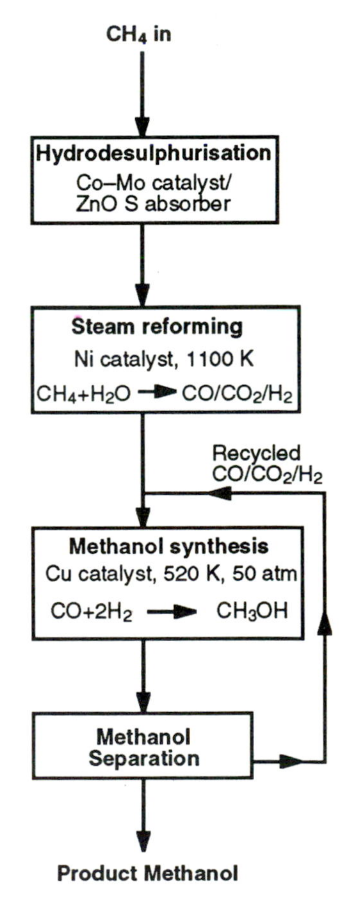

Fig. 7.3 The various processes used for conversion of natural gas to methanol.

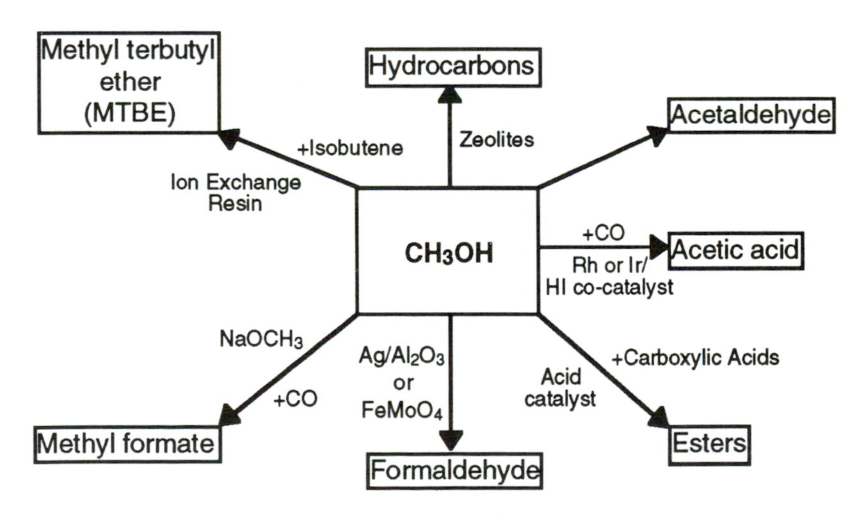

Fig. 7.4 Catalytic conversion of methanol to other products.

Formaldehyde is produced over very low area Ag, or mixed oxide catalysts ($FeMoO_4$, for instance) in a reaction known as oxidative dehydrogenation (Eqn 7.5). Acetic acid is produced by carbonylation of methanol in a liquid phase homogeneous process using Rh or Ir/Ru organometallic catalysts with HI/CH_3I as a co-catalyst.

$$CH_3OH + \tfrac{1}{2}O_2 \rightarrow H_2CO + H_2O \tag{7.5}$$

$$CH_3OH + CO \rightarrow CH_3COOH \tag{7.6}$$

A plant was built in New Zealand to convert gas into fuels and this proceeds by dehydration of methanol to dimethyl ether, which then passes through a zeolite catalyst to yield a mixture of products, mainly hydrocarbons, which can then be separated to produce fuels.

Thus it can be seen that with methane as a feedstock we have all the technology that we need to supply basic chemicals to the chemicals industry and so there should be no technological difficulties with the reduction in oil supply which is coming up in future years.

From coal

The use of coal is similar to that of methane since it can produce a synthesis gas which can be used as the major reactant for other products. One difference is the stoichiometric ratio of CO to the H_2, since coal has a chemical formula of approximately CH, with much less hydrogen than methane contains.

The Fischer–Tropsh technology was invented in the 1920s and was developed up to and during World War II; it was used to produce a wide range of products. The catalyst is similar to that for ammonia synthesis and consists of promoted Fe, run at ~320 °C and 20 atm pressure, although modern plants tend to use Co based catalysts. The range of products from this process is shown in Fig. 7.5, though the distribution is somewhat different than that from oil cracking, and includes a significant yield of oxygenated hydrocarbons and methane.

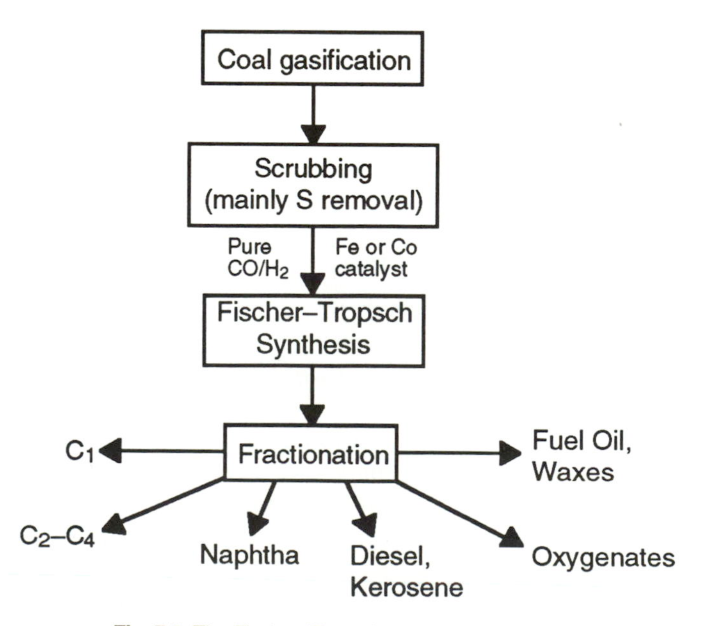

Fig. 7.5 The Fischer–Tropsch process and products.

From oil

Oil is currently the major source of petrochemicals because of its excessive cheapness (excessive because it does not yet take account of global damage inflicted by its use) and its ease of transport. As Fig. 7.6 shows, all our major industrial organic products can be produced from this source. Major outlets are, of course, fuels of various kinds for petrol and diesel engines and airplanes. Producing the correct balance of these is a tricky business and a variety of catalytic processes are used in the conversion, some are listed above in Table 7.1. For instance, in converting (reforming) a n-hexane molecule a variety of reactions can take place, as shown in Table 7.2.

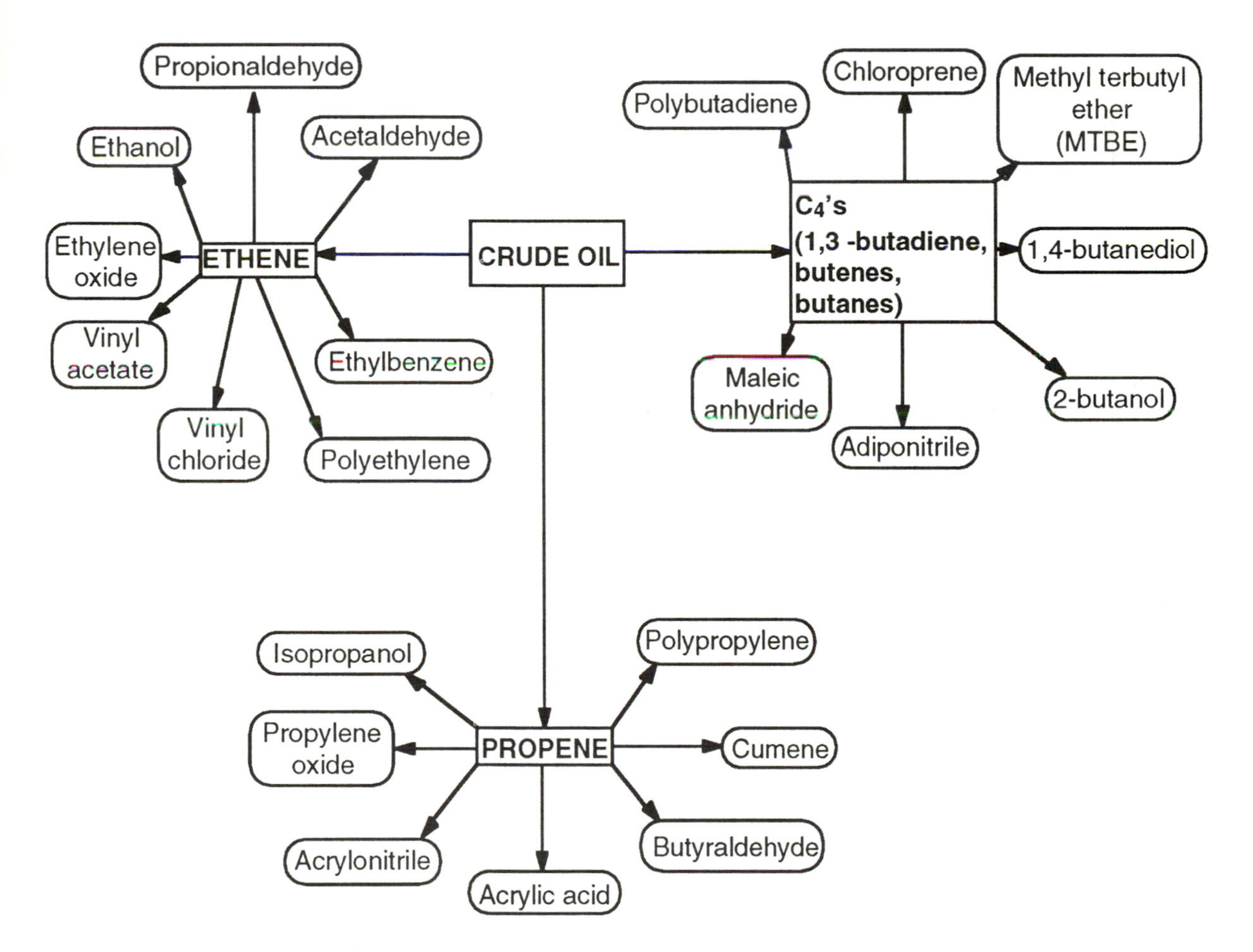

Fig. 7.6 The range of some of the major petrochemical products from crude oil.

An important class of catalysts for these conversions includes the bifunctional materials such as Pt, or Pt alloy catalysts on an acidic support (such as fluorinated η-alumina), for general reforming of a feedstock to give a higher octane rating (more aromatics and branched chains) for fuel. Such catalysts are called bifunctional because both the metallic and support components play an active role in the catalysis. The metal facilitates dehydrogenation/hydrogenation reactions and the acidic support acts on spillover dehydrogenated species to rearrange them. Hydrogen is used in the feed to keep the surface clean of cracked low molecular weight carbon species, which would quickly poison the activity of the metal function.

Table 7.2 Possible reactions of n-hexane in reforming (in the pressure of hydrogen gas)

Reaction	Product(s)	Further reactions of product
Dehydrocyclisation	Cyclohexane, methyl cyclopentane	Dehydrogenation to benzene, hydrogenolysis back to n-hexane, dehydroisomerisation to benzene
Isomerisation	i-hexane	Further isomerisation
Hydrogenolysis	Propane	Further hydrogenolysis, dehydrogenation, oligomerisation
Dehydrogenation	i-Hexene	Isomerisation, double bond migration

The ethene and propene fractions of naphtha cracking are the basis of many of our modern materials, as shown in Fig. 7.6. These are used to make plastics for mass usage (polyethylene and polypropylene) and these days can be reacted with other molecules to make specialist polymers and engineering plastics with special properties.

A large amount of ethylene is also converted to ethylene oxide and this is the largest volume selective oxidation carried out by industry. It uses a low area Ag catalyst supported on porous α-alumina (sapphire). As usual with selective oxidation reactions, the work of the catalyst is to 'beat' thermodynamics by favouring the production of ethylene oxide as opposed to combustion, which is much more exothermic. This catalysis is discussed in more detail in Chapter 9. The ethylene oxide product is used on a large scale (Fig. 7.7), as ethoxylates in washing powders, but much of it is also converted to ethylene glycol for use as antifreeze in vehicles.

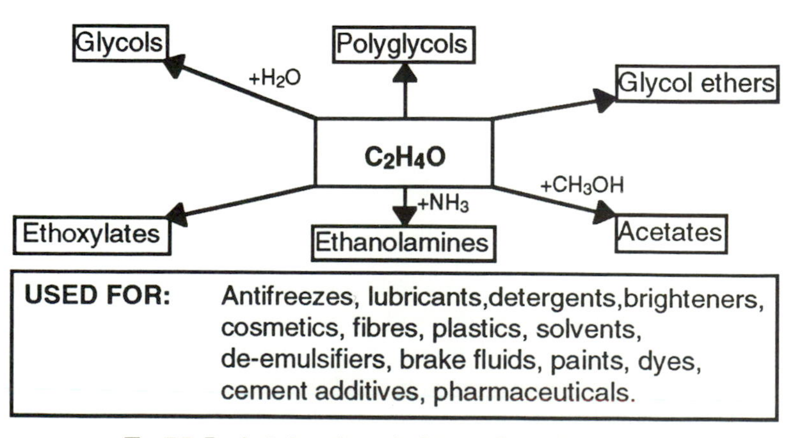

Fig. 7.7 Products from the major intermediate ethylene oxide.

Thus, a range of technologies exist to produce the major fuels and chemicals required nowadays and all three major hydrocarbon sources, solid, liquid and gas, are used in various parts of the world. Although oil is the least abundant of these feedstocks, it is the most convenient to use for a variety of reasons, including purity, ease of transport and a most appropriate C:H ratio for the main products required. However, the technology is now in place for use of the solid and gas carbon sources, for the time when the oil supply is reduced and prices of the latter become uncompetitive. A more severe problem for the future may be the huge emissions of CO_2 which the use of these carbon sources results in, and this has implications for the kind of catalysis and feedstocks used in the next century, as described further in Chapters 8 and 10.

8 Catalysis for environmental protection

8.1 Introduction

Table 8.1 Some major environmental stresses

Environmental symptom	Likely cause
Tropospheric pollution, smogs, etc.	Fossil fuel over-consumption
Acid rain	Fossil fuel over-consumption
Global warming	Fossil fuel over-consumption
Ozone hole	Chlorofluoro-carbon emissions
Decreased male fertility	Groundwater pollution

A great deal of damage has been done to the environment during the last century, mainly associated with the industrialization of society and the greater wealth and materialism of individuals. Table 8.1 lists the various areas of concern for the global environment. Many of these kinds of pollution are the result of overuse of raw materials, but much of it is due to their careless use, and inadequate attention has been paid to the by-products of these processes in the past. We have the technology in our modern world to solve all these pollution problems given the will and the backing of government. In general, commercial pressures manipulate governments to be lenient in this area; it is the advent of pressure groups, such as Greenpeace, Friends of the Earth and many others, backed by public opinion, which has forced politicians to do something about pollution. That 'something' is enacting legislation, which forces improvements in industrial and individual practices.

Legislation against pollution was initiated very early on and local laws against coal burning were enacted in European cities as early as the 13th century. More recently legislation in the 'Clean Air Act' in the UK (1956) was introduced after some particularly bad 'pea-souper' smogs in London, which were thought to have been responsible for the deaths of some 4000 people in excess of the normal mortality in 1952. Major pollutants in such smogs are particulates and SO_2, a severe lung irritant, which is acidic when mixed with moisture in the lungs, and this pollution resulted largely from the domestic consumption of coal as a fuel. As a result 'smokeless zones' were defined in which only cleaner fuels, such as natural gas, could be burned. More recently California initiated legislation to reduce the emissions of CO, NO_x and hydrocarbons from vehicles after smogs formed in Los Angeles which were thought to be due mainly to these pollutants and their photochemical products. This legislation quickly spread across the USA and more recently around the world, and now most developed countries have restrictions on vehicle emissions. These are classic examples of good leads by politicians via legislation, something which occurs, alas, too infrequently, largely due to dishonesty in some governments induced by donations from interested parties (such as elements of the road building lobby for instance).

Pollution comes from many sources as indicated in Table 8.1, and has many consequences on nature, mainly to weaken it. Nature, of course, includes human beings. Human beings are weakened in our modern society, just two of many examples of this being the huge increase in asthma in Western society, caused by a variety of environmental factors, but mainly by pollution, and the big decrease in fertility (sperm count) of males in the last 30 years.

If there is an economic push (in this case by force of legislation) then scientists can solve most problems. In the first example of catalysis used for environmental protection given in the next section, the reduction of emissions from petrol vehicles will be described. Catalysis, however, can be generally applied to environmental improvement, for the following reasons:

(a) Processes can be improved in selectivity by the application of catalysts (less by-products).

(b) They can be improved in efficiency by lower temperature operation. In the case of oxidation with air this results in less NO_x emission because NO_x formation results from direct N_2/O_2 reaction at high temperatures.

(c) Catalysts can be applied at the 'back-end' of all processes to remove a significant fraction of the pollutants by further conversion.

8.2 Mobile sources: removing pollutants from exhausts

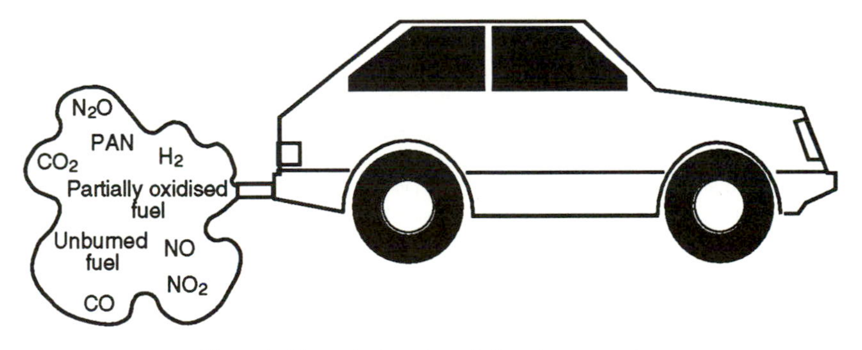

Fig. 8.1 Atmospheric pollutants from the car engine.

The problem

Figure 8.1 shows the problem with the automobile; it emits noxious chemicals, the major pollutants being CO, NO_x and hydrocarbons. Reactions in photochemical smog of various of these pollutants produce even more toxic materials (such as peroxyacetyl nitrate, PAN, which is toxic to the lungs in parts per billion). Of these pollutants the automobile is the major source of CO, and contributes ~25% to total NO_x emissions. This, combined with the huge increase in car numbers around the globe in the last 30 years, has put a heavy strain on the environment. Besides the direct human problems caused by such pollution, it is also a source of SO_2 (in small amounts compared with power-stations) and NO_x which are responsible for the production of acid rain and subsequent destruction of forests and plant life, which, in turn, has knock on effects on higher lifeforms. Figure 8.2 shows an example of the percentage of pollutants emitted by a car engine. The engine usually operates close to the stoichiometric ratio of fuel to air, at which point there is very high production of hydrocarbons, CO and NO_x. Furthermore, Pb compounds were added to fuel as 'anti-knock' agents, at a level which resulted in several kilograms of lead being emitted into the environment from each engine every year.

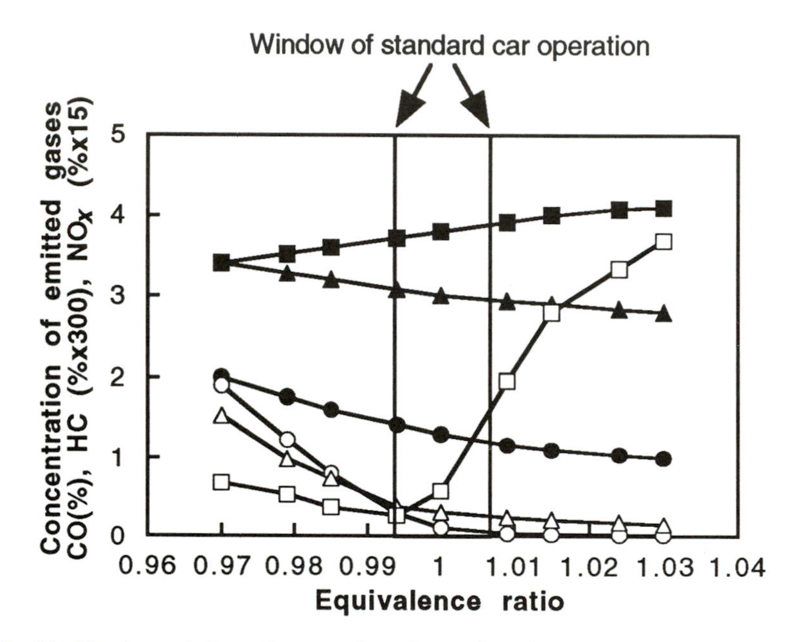

Fig. 8.2 Showing emissions of some major pollutants from the car engine (solid data points) and their reduction by addition of a catalytic converter (open points). The equivalence ratio is related to the air : fuel ratio and the limits within which the car operates are shown. Circles are for CO, triangles for hydrocarbons, squares for NO.

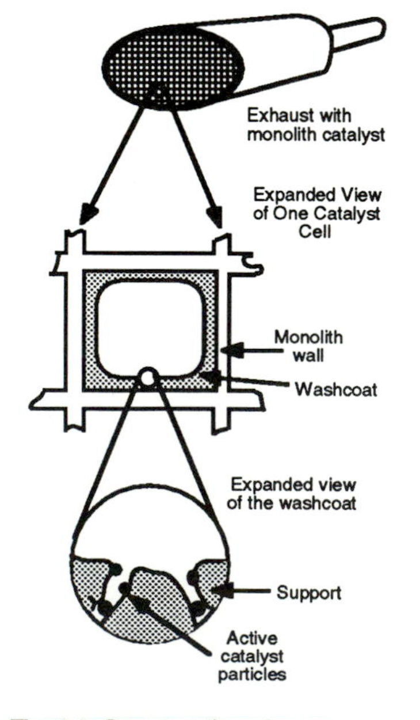

Fig. 8.3 Cross-section of a car catalyst with expanded views of the washcoat and active phase.

The solution

The solution to this problem is to remove the pollutants from the exhaust, and this is achieved by catalytic conversion. As Fig. 8.2 shows, high conversion of the main pollutants is achieved close to the stoichiometric ratio by placing a three-way catalyst in the exhaust line of the vehicle.

The catalyst used is a complex one, which fulfils a variety of functions. It is constructed as shown in Fig. 8.3. The framework of the converter is a ceramic monolith constructed of a material called cordierite, which is a Ca/Mg aluminate. It is designed to be strong and both impact and thermal shock resistant. This is very important since the converter can change temperature on starting an engine by 500 K in a short time interval. The monolith has to hold onto the catalyst through this temperature regime and so has a low expansion coefficient. The active phase has a high surface area for optimum activity and so a high area alumina support is deposited onto the monolith by a technique known as washcoating. The whole catalyst monolith is then impregnated with a solution of salts of the main active metals, Pt, Pd and Rh. Modern catalysts also have a number of other additives such as ceria, and a variety of other oxides, often present as oxygen storage media to maintain high oxidation in the fuel-rich part of the engine cycle.

This kind of system is called a three-way catalyst because it can successfully convert the three main pollutants (CO, NO_x and hydrocarbons) as follows:

$$CO + \tfrac{1}{2}O_2 \rightarrow CO_2 \qquad (8.1)$$

$$\text{hydrocarbons} + O_2 \rightarrow CO_2 + H_2O \qquad (8.2)$$

These reactions are mainly carried out by Pt which is an excellent oxidation catalyst. However, NO conversion is more difficult, since it has to be reduced, and this is achieved by a number of reactions, the most notable being those shown below. Rh is an excellent metal for these and minimizes side reactions which produce, for instance, N_2O.

$$CO + NO \rightarrow CO_2 + \tfrac{1}{2}N_2 \qquad (8.3)$$

$$H_2 + NO \rightarrow H_2O + \tfrac{1}{2}N_2 \qquad (8.4)$$

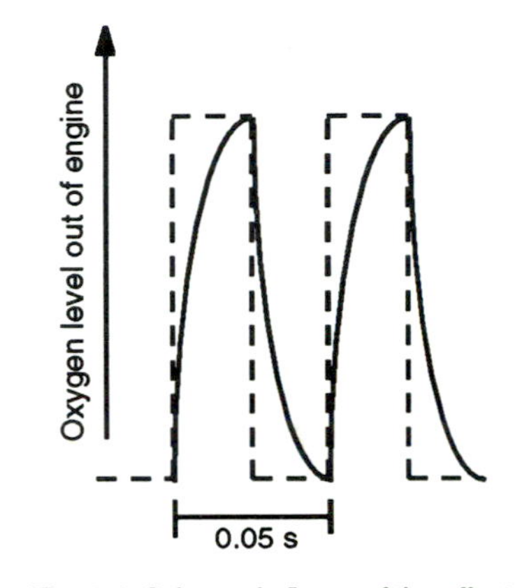

Fig. 8.4 Schematic figure of the effect of an oxygen storage oxide on the effective oxygen level in the catalyst. Dashed line, without, solid line, with the storage medium.

Other components in the catalyst play a multiplicity of roles, but perhaps the most important is in the oxygen storage capacity which has the effect shown in Fig. 8.4. The storage component takes up oxygen onto the surface during the lean cycle, and gives it out during the rich cycle, thus ameliorating the disastrous loss of NO_x conversion which would occur in the O_2-rich part of the cycle.

Legislation is continually driving technology to reduce the damaging emissions from cars, so that current European legislation dictates for the emission levels shown in Table 8.2. It should be noted that a great deal of research work is now devoted to the phenomenon of 'light-off' (Fig. 8.5), the point in temperature at which the catalyst begins to work. As Fig. 8.6 shows, the reason for this effort is clear; most of the emissions detected during a test come during 'cold start', after the engine is first turned on and before the catalyst has warmed up sufficiently to start working. If this light-off temperature can be reduced significantly, then it will have a major impact on the total levels of pollutants that are emitted from the vehicle.

Table 8.2 European limits on gasoline car emissions

Pollutant	1996 (g km^{-1})	2000 (to be finalised g km^{-1})
CO	2.2	1.0
Hydrocarbons	0.5 (Hydrocarbons + NO_x)	0.1
NO_x		0.8

Other areas for current work for reduced vehicle emissions are several fold. Some modern cars are of the lean-burn type, that is, they operate on the fuel lean, oxygen-rich side of the stoichiometric ratio (right hand side of Fig. 8.2). As can be seen in Fig. 8.2, however, conversion of NO_x is very poor under these conditions for normal three-way catalysts. Thus a new generation of catalysts has had to be designed to cope in this environment which reduce NO_x under oxidising conditions. Several classes of catalyst have been devised for this kind of operation. For instance, copper supported on a zeolite material shows very good initial activity for the reaction, but degrades in performance rather quickly in the real exhaust conditions of high temperature and humidity.

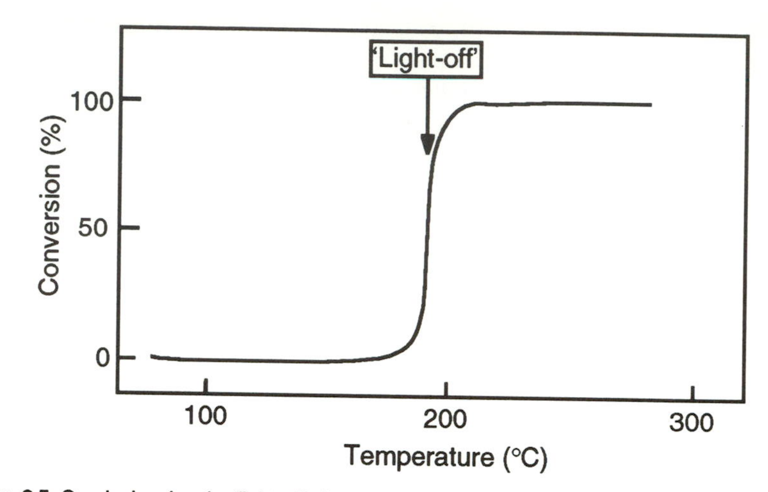

Fig. 8.5 Graph showing the 'light-off' phenomenon where conversion changes from a very low level to a very high level over a narrow temperature range. In this case light-off is at ~190 °C.

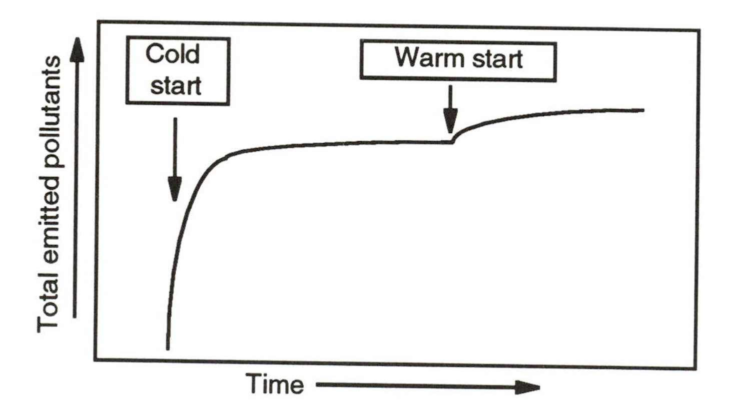

Fig 8.6 Graph showing the profile of total emitted pollutants for a car fitted with a catalytic converter. This shows that the vast majority of pollutants are emitted just after 'cold start' when the engine and catalyst are cold and the latter is ineffective. A 'warm start' occurs after the engine has been shut off for only a short time.

Of rather more promise are catalysts which have the ability to store NO_x during running and this is then removed by injection of a reductant for a short duration into the gas phase (Fig. 8.7). Such a system has recently been announced by Toyota and uses Ba as the main NO_x storage cation.

Another area of considerable research activity is the clean-up of diesel emissions, which are characterised by a high level of NO_x and particulates. The latter are soot and sulphate particles, which may be in a highly reactive form, like supported catalysts, with metals such as V and Ni deposited on the surface, these coming mainly from natural components of the oils and from additives used in engine lubrication.

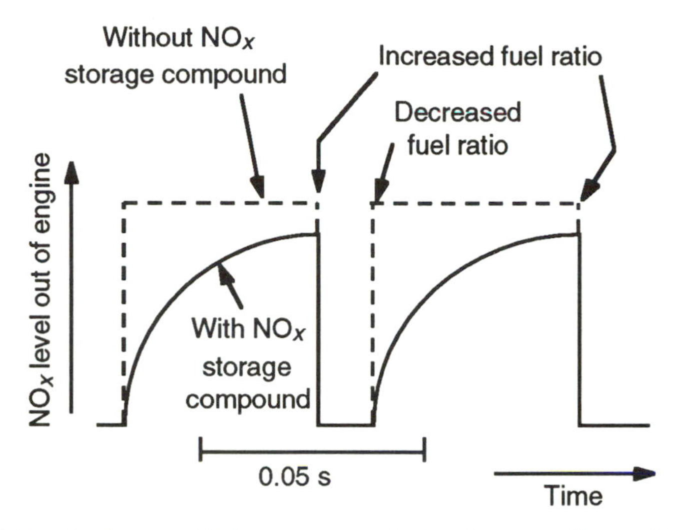

Fig. 8.7 Showing increased NO_x conversion under lean-burn conditions, with periodic increases in fuel ratio, where a NO_x storage device is used.

This is a very difficult problem since such particulates could easily block a ceramic monolith. Thus, a catalytic system for diesel clean-up would have to consist of at least two parts, as shown in Fig. 8.8, namely, a soot trap and oxidation catalyst, with a following NO_x conversion catalyst. Furthermore, all of this has to work at the low temperatures of a diesel exhaust (only ~200 °C), and, for application in most of the world, has to be sulphur tolerant. Such a system has yet to be developed for commercial use, and it must be said that an easier solution could be to eliminate diesel engines.

There are presently moves in California to demand zero-emission vehicles (ZEVs) and 10% of all vehicles sold in California will have to be of this type in the near future; ZEVs are generally electrically powered vehicles. A great deal of effort is going into this area by car manufacturers and the performance of electric vehicles (range and speed) is improving continually.

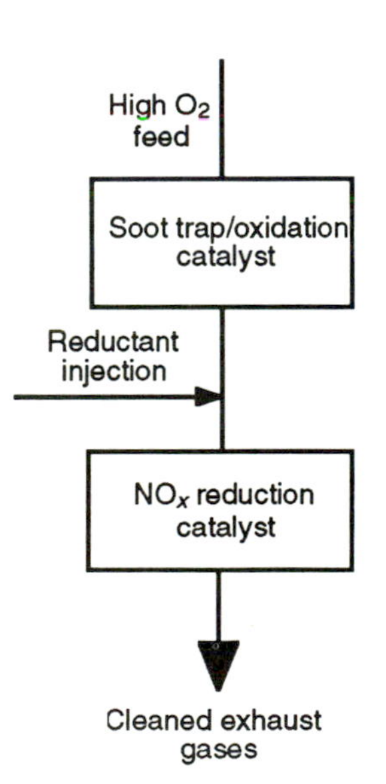

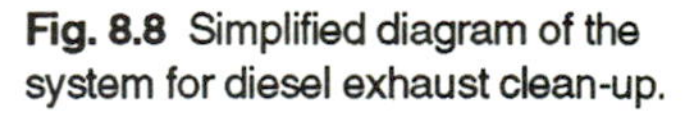
Fig. 8.8 Simplified diagram of the system for diesel exhaust clean-up.

8.3 Static sources: effluent clean-up

Static sources of pollutants are many and varied and of greatest concern are those which emit gases and liquids. In the former category are power stations (major NO_x emitters) and industrial operations, while many small and large industries have used rivers as free diluents and transport devices for liquid effluents. Again, legislation is driving for reduced emissions by both methods, especially the latter, which can be easily monitored and regulated.

Figure 8.9 gives an idealized view of a catalytic clean-up system which consists of a catalyst on the exit of the source. The catalyst may be able to do most of the work on its own, especially for oxygen-rich effluent streams in which the undesired emissions can simply be removed by catalytic oxidation. However, more commonly, oxygen or other reactants will need to be fed into the stream to depollute the emissions reactively. In some cases the added reagent itself may be a pollutant and so careful monitoring of the exit stream is required in order to ensure stoichiometric reaction and full conversion. This then requires a fast feedback mechanism to alter the ratio of reactant injected into the effluent line and this is illustrated in Fig. 8.8 and requires the addition of sensors with high molecular sensitivity and selectivity.

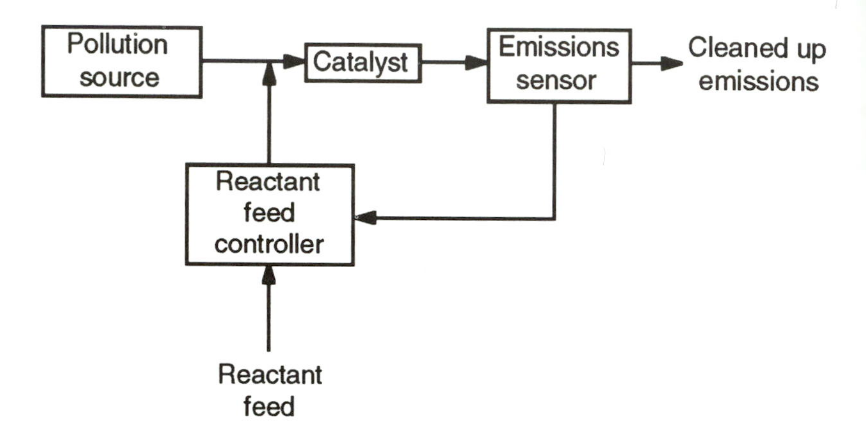

Fig. 8.9 Catalytic reactive clean-up system.

In what follows just two examples will be given of catalytic solutions to specific pollution problems.

(a) **Liquid phase hypochlorite removal.** Many aqueous effluent streams from a variety of sources produce hypochlorite (ClO^-), especially from the production and use of chlorine and chlorinated products. An elegant and simple catalytic process for its conversion has been developed which produces the following conversion with high efficiency and selectivity. This reaction,

$$2NaOCl \rightarrow 2NaCl + O_2 \tag{8.5}$$

has been shown to be tenable using catalysis and modern industrial catalysts can be of various kinds, for instance supported promoted Ni or Co catalysts used in either slurry bed or fixed bed reactors.

(b) **Gas phase NO_x removal.** A major route to NO_x removal is by reaction with oxygen and ammonia.

$$4NH_3 + 4NO + O_2 \rightarrow 6H_2O + 4N_2 \tag{8.6}$$

This kind of clean-up method is used on both the small-scale (e.g. industrial boilers) and the large-scale (e.g. nitric acid plants, waste incinerators, power plants). The catalysts used for such conversions are various, but, for instance, supported vanadium pentoxide makes a stable, active material.

8.4 Protecting the home environment: the catalytic toilet

A fascinating and unusual application of catalysis has recently been developed by Dr. Haruta and co-workers in Japan. There are circumstances in which lavatories are very small cubicles with poor air circulation, with the result that unpleasant odours can be emitted from the toilet itself, and can perhaps remain in the local environment for some time. The group in Japan have developed extremely active catalysts for oxidation of the odorous compounds (mainly amines) within the toilet at ambient temperature. The system is drawn schematically in Fig. 8.10a and the catalyst pack is shown in Fig. 8.10b. It consists of several parts, but the catalyst is gold/iron oxide supported on a zeolite, and the whole is made on a paper monolith framework. This pack weighs only 40 grams and fits into the toilet air ventilation stream above the waterline. The catalyst can last for 7 years before needing to be replaced.

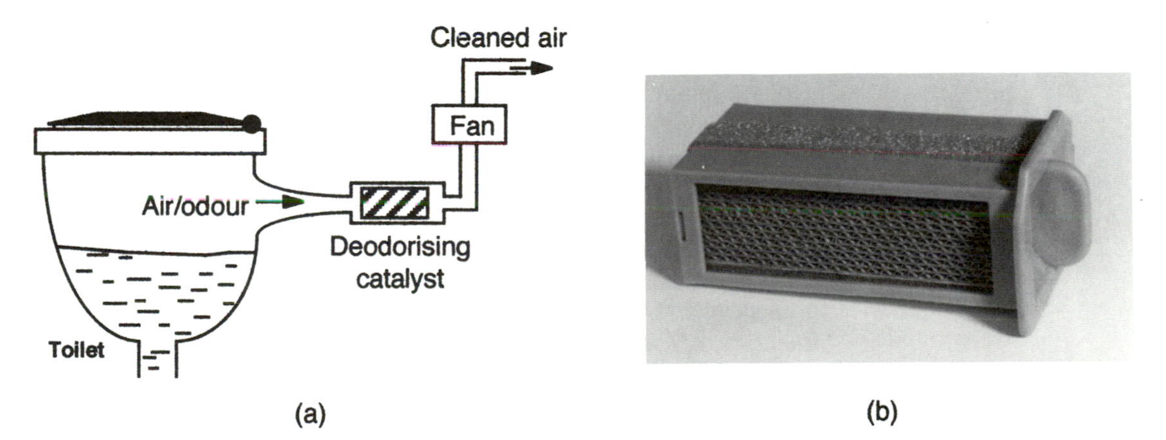

Fig. 8.10 (a) Simplified diagram of the catalytic toilet. (b) Photograph of the catalyst monolith insert for use in the toilet. [Courtesy of Dr. M. Haruta: See M. Haruta, A. Ueda, S.T. Subata and R.M. Torres, Catal. Today 29 (1996) 443; and T. Matsumoto and K. Tabata, Nat. Tech. Report 40 (1994)32.]

The development of this system is a marvellous achievement of catalytic science (even if mundane in application), all the more impressive because of its use of a metal, gold, hitherto widely considered to be useless in the catalytic context. This further presages the advent of catalysis more widely into the domestic scene as discussed in Chapter 10.

9 Catalysis in everyday life

9.1 Introduction

In the previous chapters we have concerned ourselves with a range of catalytic principles and some examples of industrial catalysis. In this chapter some examples of catalysis which impinge rather directly on individual modern lives will be given. As already mentioned, society is highly dependent on the chemical industry and our modern lifestyle would be impossible without this business. The vast majority of chemical products are formed using catalysis.

One of the most directly evident everyday uses of catalysis has been given above – namely automobile catalysis. This is perhaps the catalytic technology of which people are most aware nowadays. However, there is a wide range of catalytic processes which are used to make everyday products, some of which are a surprise, and in the following sections some of the larger scale of these will be described.

9.2 Food from gas: fertilizer production

Most modern western economies are dependent on livestock for food production and cows are fed largely from grass and grains. For the maximum intensity of production (hamburgers per square metre) fertilizer has to be added to fields to increase grass growth rates and to increase grain yields (approximately 10 kg of grain are required per kg of beef produced). Cows are vegetarian and require such non–meat inputs; feeding alternative meat-based sources has led to the BSE ('mad cow disease') scare in the UK. The yield of grass and grain used for feed is improved by the addition of nitrogen to the soil in the form of ammonia-based fertilizers (e.g. NH_4NO_3).

The ammonia is produced by the Haber process for nitrogen fixation which is exothermic.

$$N_2 + 3H_2 \rightarrow 2NH_3 \tag{9.1}$$

Since there is also a pressure drop upon synthesis, the thermodynamics dictate that the reaction should be optimised at low temperature and high pressure. At very low temperatures the reaction becomes kinetically limited and so industrially temperatures of around 450 °C and pressures of 100 atm are used, under which conditions yields of NH_3 are ~14%, close to the equilibrium value. Because conversion is low the ammonia is trapped out in a cooler/refrigeration unit at the exit of the reactor and the unconverted N_2/H_2 is recycled to the front of the reactor. The catalyst used is mainly iron (Table 9.1) with small amounts of a promoter (potash) and a support (alumina),

Table 9.1 Approximate composition of ammonia synthesis catalysts. Exact levels vary with manufacturer

Component	Level (%)
Fe_3O_4	96
K_2O	0.5
CuO	2
MgO	0.2
Al_2O_3	2
SiO_2	0.5

which also plays some role in the activity. This is a very strong and stable catalyst.

Although the potash and alumina are present in low amounts in the bulk, various surface science studies have shown that these are preferentially segregated to the surface in the working catalyst. Thus, although at <1% in the bulk, potash has a surface coverage of ~10% (Fig. 9.1). In order to produce the N_2 and H_2 for the synthesis, however, a number of other catalysts have to be used, as illustrated in Fig. 9.2. These catalysts are required for several steps in the conversion of natural gas (methane) into ammonia.

(i) Hydrodesulphurisation. Removal of sulphurous stench agents added by the gas companies to make the gas detectable to the nose. This removal is achieved by conversion to H_2S and then absorption/reaction of this with ZnO.

(ii) Steam reforming. As described in Chapter 7, using a Ni catalyst.

(iii) Water gas shift reaction. To convert CO into CO_2 by reduction of water to produce more hydrogen. Carried out in two stages for kinetic and thermodynamic reasons.

(iv) Methanation. Removes the small amounts of remaining CO by hydrogenation back to methane. Although the CO level is low, it would be sufficient to block up the surface sites of the iron catalyst and quickly poison activity.

(v) Ammonia synthesis. Mainly Fe catalyst as described above, which hydrogenates nitrogen, in a very pure feed gas.

(vi) Ammonia oxidation. This produces NO which is further oxidised and hydrated to produce nitric acid. This is then reacted with ammonia to make the final product fertilizer.

Although some recent advances have been made (Ru catalysts) all plants around the world use catalysts essentially the same as those developed by BASF at the start of this century.

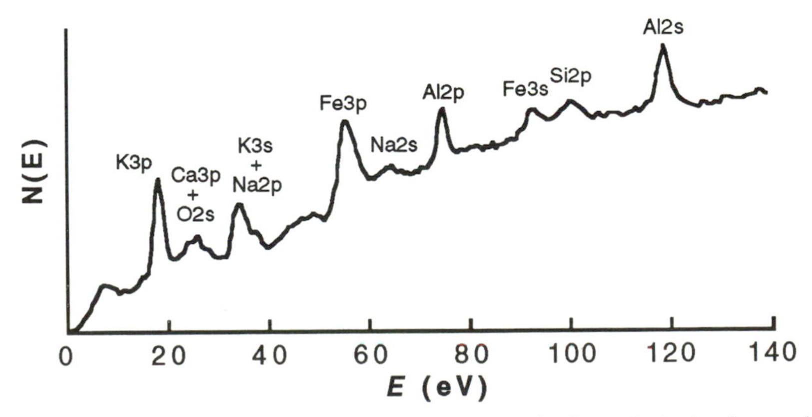

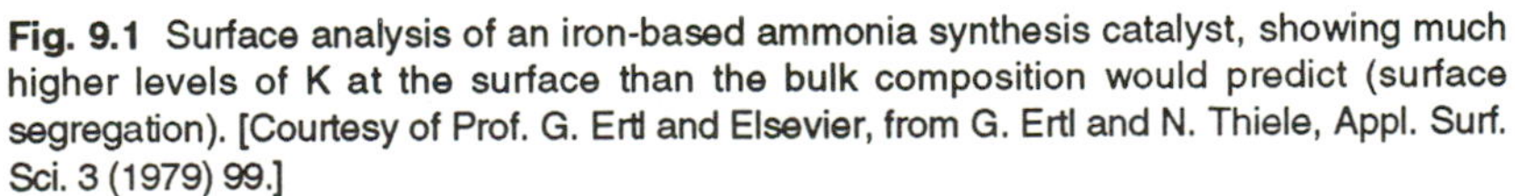

Fig. 9.1 Surface analysis of an iron-based ammonia synthesis catalyst, showing much higher levels of K at the surface than the bulk composition would predict (surface segregation). [Courtesy of Prof. G. Ertl and Elsevier, from G. Ertl and N. Thiele, Appl. Surf. Sci. 3 (1979) 99.]

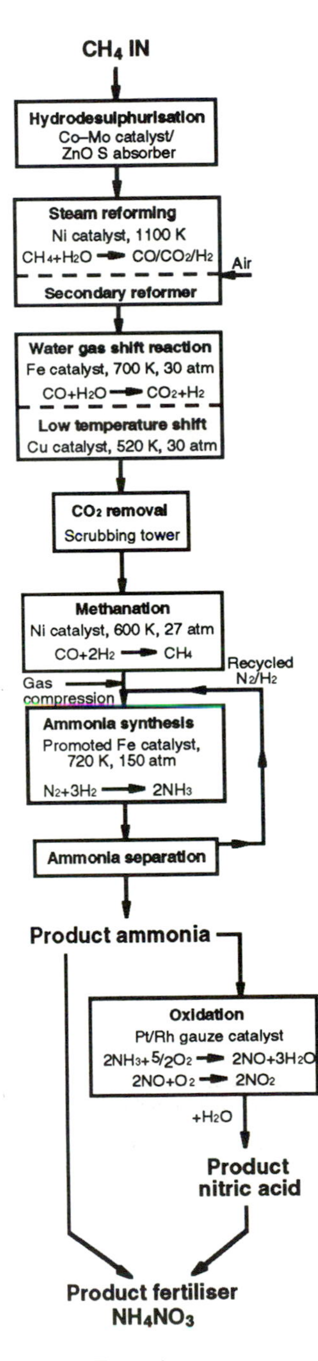

Fig. 9.2 The various processes required to produce fertilizer.

9.3 Modifying food: margarine production by hydrogenating vegetable oil

Vegetable oil is mainly composed of unsaturated fats such as linolenic and linoleic acids. These are liquid oils which are not pleasant for use on bread, they tend to degrade easily and are often not convenient for baking. Hydrogenation of one or more of the double bonds in the molecules increases saturation (mono-unsaturated oleic acid is the main product), and increases hardness and 'spreadability', thereby producing the margarine familiar to everyone.

This saturation is achieved using heterogeneous catalysis, with Ni-based catalysts supported on silica, the reaction being run at 420 K at c. 3 atm pressure in batch mode, with subsequent removal of the catalyst by filtration.

9.4 Vinegar from gas: homogeneous catalysis on the grand scale

The majority of large-scale catalytic processes are heterogeneous, most often gas–solid reactions, sometimes liquid–solid, but here we will present one particular process which is run on the scale of up to 500,000 tonne/year in one plant, using a homogeneous catalytic process.

Acetic acid is produced by the carbonylation of methanol, the reactants themselves being produced from natural gas. Acetic acid from this industrial source is used widely by the food industry, though the large majority of the product is used to make acetates, especially vinyl acetate.

$$CH_3OH + CO \rightarrow CH_3COOH \tag{9.2}$$

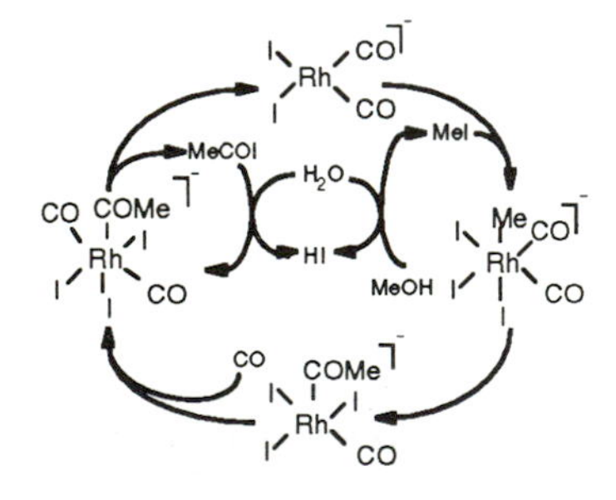

Fig. 9.3 The cycle of reactions involved in methanol carbonylation. [From original work of T. Dekleva and D. Forster, Adv. Catalysis 34 (1986) 81). See also P. Maitlis, A. Haynes, G. Sunley and M.J. Howard, J. Chem. Soc., Dalton Trans. (1996) 2187.]

In this case the reaction is carried out by a dissolved monomeric organometallic catalyst which passes through several intermediate forms, though the predominant species observed in solution is $[Rh(CO)_2I_2]^-$, since the reaction is run in the presence of a co-catalyst, a mix of HI/CH_3I. The mixture is approximately 10^{-3} M in Rh and is run at around 180 °C and 35 atm pressure in a continuous process. The catalytic cycle is thought to proceed in the manner shown in Fig. 9.3, with several organometallic intermediates, some short-lived and some detectable at a steady concentration in the reacting medium. Recently, newly developed catalysts have been announced, based on Ru promoted Ir organometallics of a similar type. These have better stability characteristics, are already in use in plants and may supplant the Rh-based process.

9.5 Furniture from gas: formaldehyde production

Methanol is produced from natural gas as described in Chapter 7. A great deal of the methanol manufactured in this way is subsequently converted into formaldehyde as follows

$$CH_3OH \rightarrow H_2CO + H_2 \tag{9.3}$$
$$\Delta H = +86 \text{ kJ mol}^{-1}$$

$$CH_3OH + O_2 \rightarrow H_2CO + H_2O \qquad (9.4)$$
$$\Delta H = -156 \text{ kJ mol}^{-1}$$

and, in turn, much of the formaldehyde is converted into the resins which glue wood chip particles together to make furniture board. The world production is estimated at 4 million tonnes. The reaction above (9.3) is dehydrogenation, and is endothermic, whereas reaction 9.4 is exothermic and is called an oxidative dehydrogenation. The latter is more favourable, but combustion (Eqn 9.5) is even more favourable

$$CH_3OH + \tfrac{3}{2}O_2 \rightarrow CO_2 + 2H_2O \qquad (9.5)$$

and so to avoid over-oxidation the reaction is run in a substoichiometric amount of oxygen, essentially a mix of reactions 9.3 and 9.4.

The preferred catalyst used for this reaction has been Ag, in the form of powder or sponge, of low surface area, to minimise combustion. The reaction proceeds through a methoxy intermediate which decomposes to yield formaldehyde, whereas the combustion route occurs through over-oxidation of this intermediate through a formate intermediate which decomposes to give CO_2 and H_2, the latter reacting with surface oxygen in a facile manner to produce water (Fig. 9.4). This process is run at temperatures around 900 K and pressures just above atmospheric. To help to avoid combustion the catalyst is of low surface area and the reaction is run with short residence times over a narrow plug of catalyst. Recently, new catalysts have been introduced and the selective oxidation can be carried using oxide catalysts such as iron molybdate, $Fe_2(MoO_4)_3$. The advantage of these catalysts is that complete oxidation is less easy, and so the reaction can be run in higher oxygen levels. The reaction is then more exothermic and is run at lower temperatures, of ~650 K. In fact, high levels of air are used to minimise flammability risks. The plant is of the multitubular reactor design with a heat transfer medium flowing between the tubes.

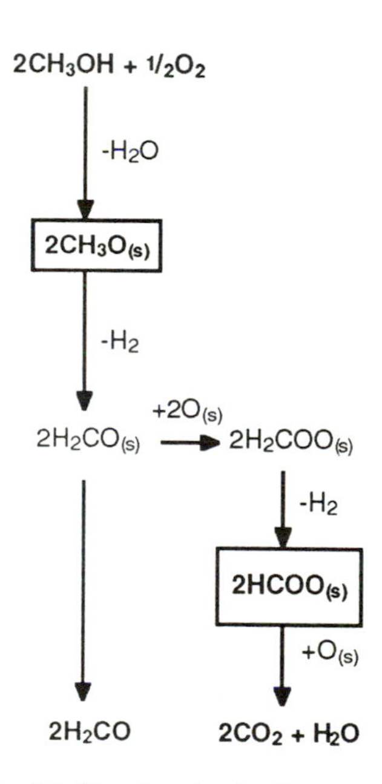

Fig. 9.4 The steps involved in heterogeneous methanol oxidation, showing the surface methoxy as the major intermediate in the selective oxidation route, and formate in the combustion route.

9.6 Clothes from oil: fibres and plastics

All plastics which are produced from feedstock oil use catalysis in their production. Modern society is now built on these materials, quite literally, with an enormous expansion in the use of plastic materials in the building sector, and most modern cars have high levels of plastics in their construction. So much so that in tests of emissions from cars fitted with an exhaust converter, the small amount of volatile organics in plastics on a new car can contribute in a major way to the total pollution load measured! Plastics are used widely in every aspect of daily life from food wrapping and drinks containers (usually PET, polyethylene terephthalate), to general packaging, to clothes. The vast majority of these materials are made from oil and the bulk plastics come mainly from the catalytic conversion of ethylene and propylene. If we use the former as an example, Ziegler and Natta won the Nobel Prize for chemistry for their reaction of a then new class of catalysts,

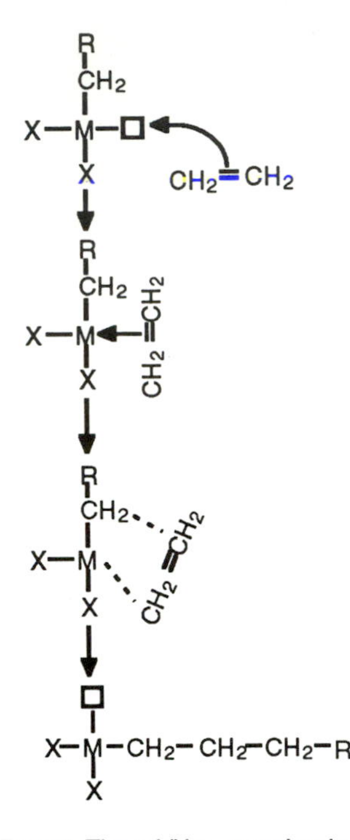

Fig. 9.5 The addition steps involved in polymerisation and chain growth on a Ziegler–Natta catalyst.

based on titanium chloride/aluminium alkyl, which are capable of polymerising a wide range of such reactions. They appear to work at an active centre in titanium where the alkene can π bond to a vacant coordinate site and chain lengthening comes from addition of groups from an adjacent site on the metal centre (see Fig. 9.5), essentially by insertion in the growing chain. The catalysis is heterogeneous in that the catalyst is a solid and the reactant is in the gas phase, but the product is a solid with the catalyst intimately locked into it. The catalyst is very efficient, however, and represents a very small fraction by weight of the final polymer. Thus, the catalyst has to be continually added to the process. New, more efficient and more specific catalysts for these processes are continually being developed.

There are a range of different processes for production of different plastics and sometimes this is very involved, requiring multistep processing. An example of this is nylon production, which involves approximately 13 catalytic steps (some routes involve more catalytic steps) to produce it from the original starting materials of oil (for the organics involved, i.e. benzene, propene) and natural gas/air/water (for ammonia production, as described in Section 9.2 above). The overall steps involved are shown in Fig. 9.6 below.

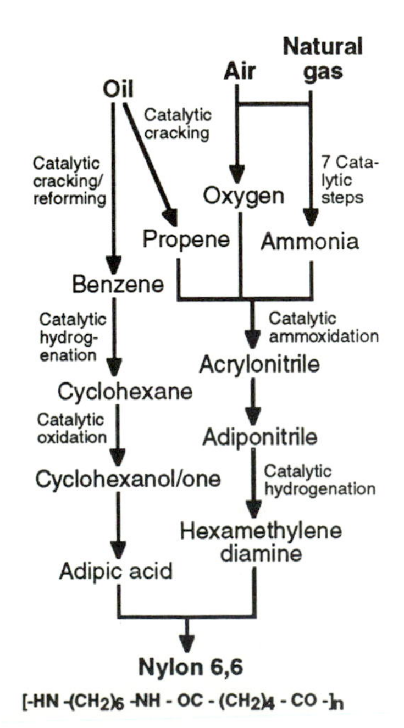

Fig. 9.6 The steps involved in nylon production from basic materials.

9.7 Washing powder additives: selective oxidation of ethylene

Ethylene oxide is used for a wide variety of products (see Fig. 7.7) and its production represents the largest tonnage selective oxidation process in use. A supported Ag catalyst is employed and works at approximately 80% selectivity. In fact, supported Ag alone yields ethylene oxide but only with about 50% selectivity. Current commercial catalysts have at least one promoter (such as an alkali oxide) doped onto the catalyst surface, and a selectivity enhancer (a chloro-organic such as vinyl chloride) is added to the gas feed at the ppm level. The latter is the most important of these modifications to the process, increasing the selectivity up to ~ 70% or more and there is an anecdote to describe its discovery. An ethylene oxide plant periodically improved its selectivity markedly and it was eventually realised that this improvement corresponded with windy days on which the wind blew from a particular direction. This direction had a chlorine plant upstream of the ethylene oxide plant, and small amounts of chlorine effluent got into the air intake of the oxidation process causing temporary improvement in ethylene oxide yield and selectivity.

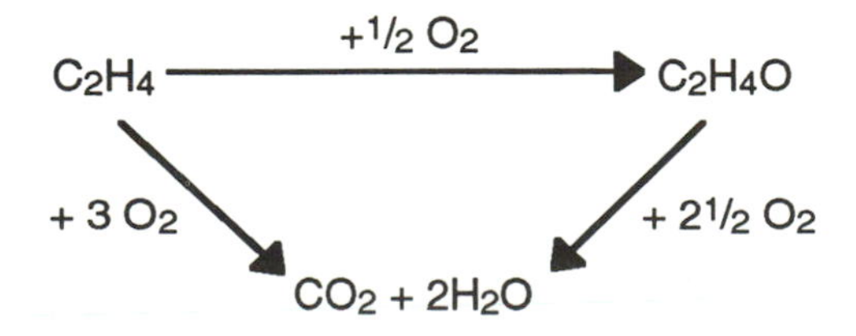

The reactor for this process is a multitubular one, usually cooled by boiling kerosene, and operates at ~250 °C. The cooling system needs to be very efficient because of the large heat output of the non-selective combustion part of the process. Operating pressures are usually ~12 bar.

The biggest use of ethylene oxide is in its glycol form as antifreeze in cars, but it is also used in washing powder as ethoxylates, and in the making of a wide range of other products, such as cosmetics, fibres and pharmaceuticals (see Fig. 7.7).

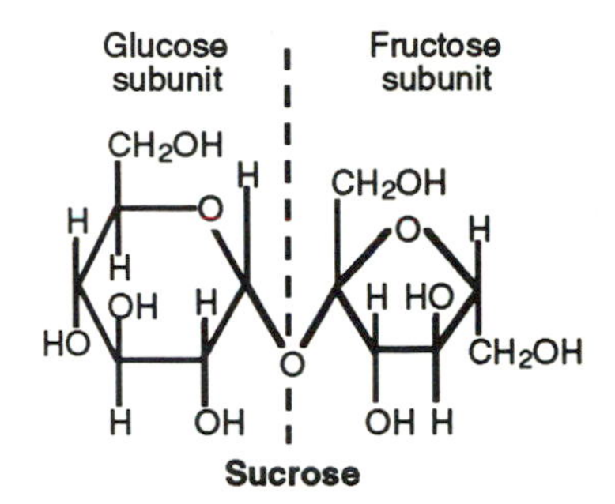

Fig. 9.7 The chemical structure of sugars.

9.8 Sweet drinks from corn: immobilised enzyme catalysts

Sucrose (Fig. 9.7) is a disaccharide that has been traditionally used as a sweetener, but a preferable sweetener is the one used by nature to make fruits sweet, namely fructose, which is nearly twice as sweet as sucrose per unit weight, thus enabling use of less carbohydrate to sweeten drinks and foodstuffs. In the 1970s there was an economic drive to produce this sweetener due to the high price of sugar cane. A process was developed using a natural enzyme called glucose isomerase, derived from the bacterium *Streptomyces* (also the source of the antibiotic streptomycin). The reactant is a by–product of corn milling, consisting of a syrupy solution of 95% glucose, and this is converted into a solution with approximately 50% fructose, which is much sweeter than the starting material (fructose is 2.5 times sweeter than glucose). This material is the so-called 'corn sweetener' in many drinks and foodstuffs. In order to maintain high activity and to enable its easy separation from the liquid the enzyme catalyst is supported ('immobilised') on a ceramic coated with a polymer to which the enzyme can adhere.

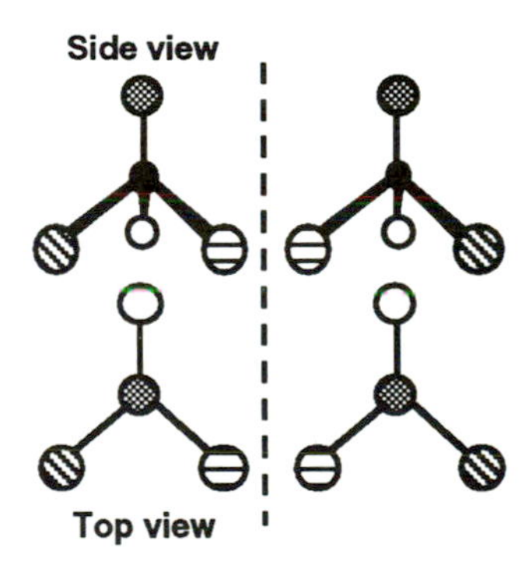

Fig. 9.8 A stereogenic centre is one which is not superimposable on its mirror image.

9.9 Catalysis for health: pharmaceuticals by asymmetric synthesis

Nature is usually very specific in the way it makes molecules, especially so when a stereogenic centre is involved (see Fig. 9.8). When there are two different molecules, it usually chooses the one which has the desired properties, in a biological sense. Different chiral versions of the same molecule often have very different properties, especially in terms of fragrance. Thus L-limonene smells of lemons, whereas D-limonene smells of oranges (Fig. 9.9). In the plants which produce these fruits, there are natural processes to make only one of the enantiomers with very high selectivity.

This is a big growth area for the application of catalysis at present. Many drugs are more efficacious in only one chiral form and so asymmetric catalysis is now being widely applied in this area. A famous example acts as a cautionary tale for this field, namely the thalidomide tragedy of the 1960s. Here the drug, as its *R*-isomer, was found to be a useful sedative, but when applied to large-scale production the racemate, an equal mixture of *R* and *S* forms was made, with the tragic consequences of the birth of many deformed

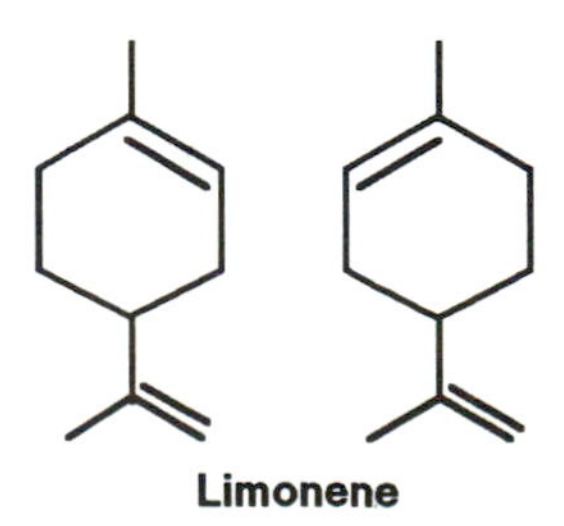

Fig. 9.9 Chiral carbon in limonene.

babies. This is because the *S*-isomer is a teratogen, which also interfers with the natural abortion function in human beings.

The ability to produce one form of isomer is therefore of great importance and interest to the pharmaceutical industry, and catalysts have a significant role to play in this development. An example of this is a catalytic route to the drug L–dopa (Fig. 9.10) used in the treatment of Parkinsons disease, although here the R–form is largely inactive, it has to be separated from the L–form, which is a costly process. Homogeneous catalysts can enable this conversion (which involves one-sided hydrogenation of an olefinic bond) with 95% selectivity. Although many of these kinds of processes are homogeneous, this involves costly operations for removing the catalyst from the liquid phase product. If heterogeneous catalysts could be prepared, then separation would become much simpler, and making the process continuous would be easier. Thus, there is a drive at present for the development of heterogeneous asymmetric catalysts, especially as a result of FDA regulations for low heavy metal concentrations in the product.

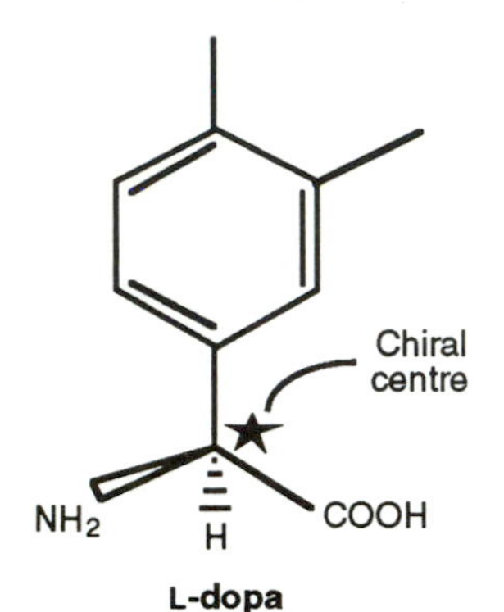

Fig. 9.10 The structure of L-dopa.

10 Catalysis for the future

10.1 Introduction

In all that has gone before in this book we have considered the fundamentals of adsorption and catalysis and their application to current processes. To conclude, the use of catalysis in the future will be considered.

The main areas of importance here are threefold: (i) new routes to fuels; (ii) new routes to commodity chemicals from alternative feedstocks; and (iii) new areas of the application of catalysis to environmental protection. However, first we will consider how the process of catalytic invention and development will change as we advance into the 21st century.

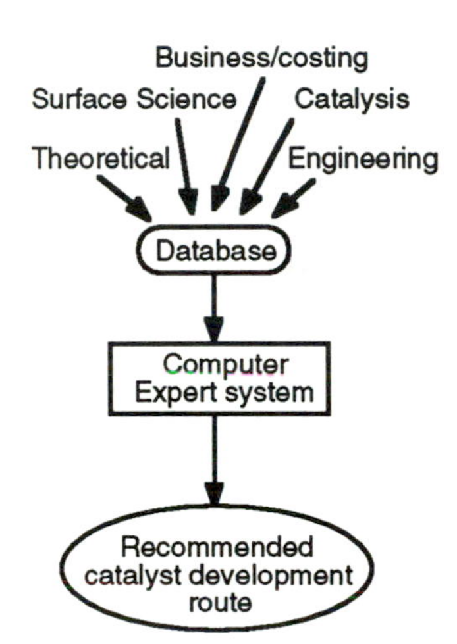

Fig. 10.1 Inputs to an 'expert system' for catalyst design.

10.2 Catalysts by design: chemistry in cyberspace

Over recent years the importance of the application of computer technology to the field of chemistry has been realised. Computers are an integral part of the scientific laboratory and office, driving experimental equipment and analysing data. They have now gone beyond this and are being used in new frontiers, in many cases to reduce the need for experimentation or to lead experiments in alternative directions. This has been achieved by the increase in personally accessible computer power as well as by the rapid development of molecular modelling. This is evidenced by the large number of specialist modelling companies now selling products to the chemical industry and academia and this has had an effect in the area of drug design in particular. Catalysis has been rather backward in this area, but in the 21st century it will mature from a 'black art' to a truly predictive scientific discipline by the use of computer-based 'expert systems' which may work broadly as shown in Fig. 10.1. This is part of the approach outlined in Section 4.3. Experimentation and testing will still be necessary but will be strongly led by such expert systems.

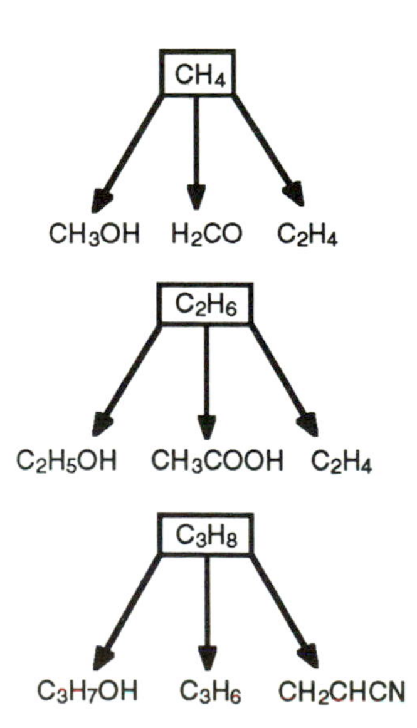

Fig. 10.2 Possible future commercial products from the direct catalytic conversion of alkanes.

10.3 Chemicals from alternative feedstocks

Many new routes to commodity chemicals are presently under consideration or development. The current use of feedstocks has been outlined in Chapter 7, oil still being the major source of chemicals and fuels. However, CH_4 and coal are present in much greater quantities on the earth and there is increasing use of these sources for a variety of applications. Their use as alternative routes to fuels and basic chemicals is restricted to a few locations in the world at present, but this is likely to expand in the medium term as the oil price increases due to diminishing production. There are also many possibilities for the conversion of the simple alkanes (Fig. 10.2), most of which are currently burned as cheap fuels, or are flared off at well-heads. In particular, several possibilities exist for the conversion of methane, including the high temperature coupling to ethene (using oxide-type catalysts initially discovered by Lunsford in Texas) or by direct oxidation to methanol and formaldehyde. Alternative routes using alkanes instead of processes currently using alkenes include, for instance, the ammoxidation of propane instead of propene, and BP America is currently developing catalysts for this process based on $VSbO_4$ as the active material, and including gas phase reaction modifiers.

10.4 The fuel of the 21st century: catalytic hydrogen production

It is likely that there will be severe curtailment of the ability of society in the coming century to continue to burn fossil fuels due to the burden of CO_2 upon the atmosphere and its contribution to global warming. This will lead to legislation that will encourage the use of non-greenhouse fuels. A major alternative is hydrogen, since, as shown in Fig. 10.3, it enables a clean cycle which, in effect, converts sunlight into power.

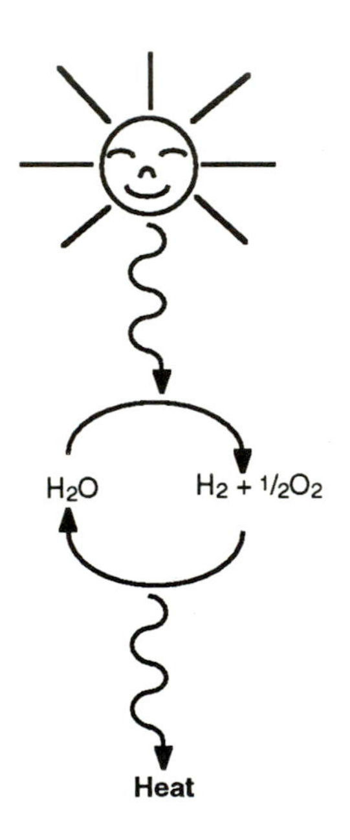

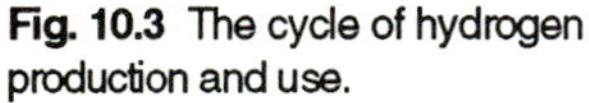

Fig. 10.3 The cycle of hydrogen production and use.

There are a variety of opportunities for the production of hydrogen in this way, for example by the combination of solar photovoltaics and electrolysis, or by electrolysis at hydroelectric plants, but there is also a possible photocatalytic process. This uses photons to provide the thermodynamic impetus to crack water and the catalysis lowers the kinetic barrier to break the OH bonds involved. Figure 10.4 shows the type of catalysts which are currently being developed for such processes.

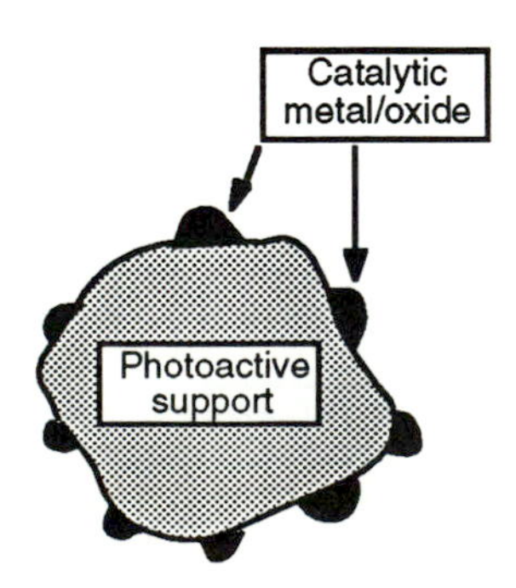

Fig. 10.4 Photoactive catalyst for water splitting to produce hydrogen and oxygen.

It is possible that in the next century hydrogen produced in this way will be used for domestic heating and cooking and that it will be used in cars. In particular, in the latter case, and as described in more detail below, fuel cells are likely to be the engines of the next century for motor vehicles. These can be powered by hydrogen, but currently it's more convenient to reform hydrocarbons (such as methanol) to produce hydrogen (and throw away the CO/CO_2 carbon end-product). Again, future legislation may prevent such engines using fossil fuels, in which case hydrogen may have to be stored on board.

10.5 New transport options: fuel cells

California has pioneered the improvements in pollution emission per car by introducing novel legislation, the most significant being that which led to the introduction of catalysts to the tail pipe of the car to reduce hydrocarbons, CO and NO_x (see Chapter 8). California has gone on to make these regulations more and more stringent, something which has kept the industrial catalytic chemists in this field busy for the last 20 years or so. This legislation has spread around the world so that we now have the situation where most developed countries have catalysts fitted to most new vehicles. However, there is still much development to do in this area, such as the application of catalytic depollution to diesel engine vehicles, to trucks and to buses, and to develop lean-burn engines which have high fuel efficiency but reduced NO_x emissions.

However, California is currently legislating for zero-emission vehicles (ZEVs). Here the emphasis will be on electrically powered vehicles, and there are battery driven cars currently available. However, these generally have a rather limited range, need regular recharging (taking considerable time) and require a heavy load of batteries to work. A more promising alternative is the electrically powered car which has on-board generation in the form of a fuel cell. This is an electrocatalytic device, shown in Fig. 10.5, which converts chemical energy (for instance hydrogen/oxygen, or methanol/oxygen) into electrical energy, forming combustion products in the process. In this sense only the H_2/O_2 powered fuel cell is truly 'zero emission', others using a fossil fuel as the original C source are CO_2 emitters.

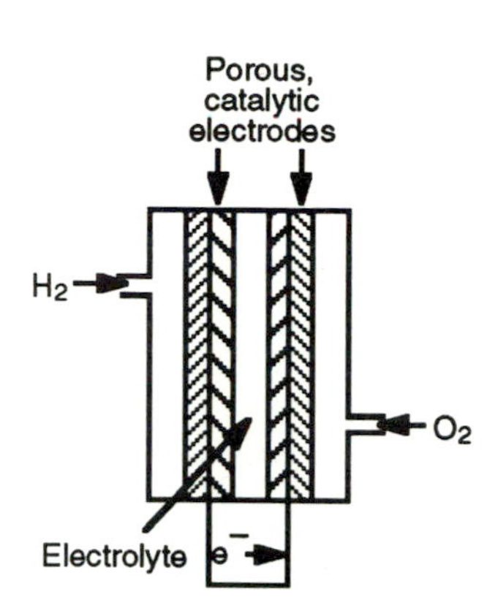

Fig. 10.5 A simplified diagram of the processes occurring in a catalytic fuel cell.

Enormous strides have been made in recent times in the development of this technology, such that it appears likely that cars powered in this way will be available early in the next century. Of crucial importance in this area is the continued development of the catalytic side of this story, especially for methanol powered fuels cells where CO is a major contamination problem, poisoning the catalytic effectiveness of the precious metal component and reducing efficiency (i.e. mpg of methanol). This is a rich area for collaborative research between those with an applied interest in such catalysis and those with an academic interest in understanding surface reactivity.

10.6 Catalysis in the home

This may well be an area for considerable expansion of catalysis, especially for the reduction of indoor pollution. Catalysts do already exist in the home – for instance, the catalytic toilet described in Section 8.4, gas powered hair dryers (using catalytic combustion of butane to produce heat) and in self-cleaning ovens. However, it is now widely recognised that people can be poisoned in the home by gaseous pollutants, either short–term (e.g. CO poisoning from incomplete combustion of methane in gas fired water heaters) or longer term from low levels of CO or formaldehyde (the latter coming from furniture materials and insulating fillers in cavity walls). These can be overcome by the use of high efficiency oxidation catalysts (such as the Au catalyst described in Section 8.4) which could be permanently mounted in the home, perhaps with a gas recycler attached so that the room air is constantly pumped through it.

10.7 Catalysis for drugs

This area was covered in Chapter 9, but is set for enormous expansion as the use of catalysis for stereospecific reactions becomes recognised. Legislation is driving for only one form of enantiomer to be present in drugs which are offered for license and for the properties of all isomers to be identified. The body often only recognises one optical isomer for a particular biochemical pathway, the other isomer may be inactive or may be involved in a completely different pathway. The major pharmaceutical companies are currently investing considerable sums of money into the development of such processes.

10.8 Conclusions

Catalysis is intimately involved in all aspects of our modern society and all organic products are based on processes using catalysts, usually heterogeneous catalysts. They are used to reduce the energy and cost requirements of the process and to reduce polluting emissions from inefficient conversion. The future of catalysis looks extremely bright, especially as it becomes more widely recognised that this technology can bring further improvements to our life. They will be more extensively applied to the development of novel materials and drugs, to more efficient processing, and to reduction of the pollution burden put upon nature by the high activity of the world's human population.

Further reading

1. G.C. Bond, Heterogeneous Catalysis (2nd edition), Oxford University Press 1987
2. I.M. Campbell, Catalysis at Surfaces, Chapman and Hall 1988
3. J. Dumesic et al, The Microkinetics of Heterogeneous Catalysis, ACS 1993
4. R. Gasser, An Introduction to Chemisorption and Catalysis by Metals, Oxford University Press 1987
5. B.C. Gates, Catalytic Chemistry, Wiley 1992
6. C.A. Heaton, An Introduction to Industrial Chemistry (2nd edition), Blackie 1991
7. D.A. King and D.P. Woodruff, The Chemical Physics of Solid Surfaces and Heterogeneous Catalysis, Elsevier, in a number of volumes from 1987 (continuing)
8. G.A. Somorjai, Introduction to Surface Chemistry and Catalysis, Wiley 1994
9. K. Tamaru, Dynamic Heterogeneous Catalysis, Academic Press 1978
10. J.M. Thomas and W.J. Thomas, Principles and Practice of Heterogeneous Catalysis, VCH 1997
11. R.A. van Santen, Theoretical Heterogeneous Catalysis, World Scientific 1991
12. R.A. van Santen and J.W. Niemantsverdriet, Chemical Kinetics and Catalysis, Plenum 1995
13. A. Zangwill, Physics at Surfaces, Cambridge University Press 1988

Oldies, but goodies

1. G.C. Bond, Catalysis by Metals, Academic Press 1962
2. J.M. Thomas and W.J. Thomas, Introduction to the Principles of Heterogeneous Catalysis, Academic Press 1967

Index

Page numbers in italics refer to figures.